EXTREME CONFLICT AND TROPICAL FORESTS

WORLD FORESTS

WFSE Series Description

As forests stay high on the global political agenda, and forest-related industries diversify, cutting edge research into the issues facing forests has become more and more transdisciplinary. With this is mind, Springer's *World Forests* series has been established to provide a key forum for research-based syntheses of globally relevant issues on the interrelations between forests, society and the environment.

The series is intended for a wide range of readers including national and international entities concerned with forest, environmental and related policy issues; advanced students and researchers; business professionals, non-governmental organizations and the environmental and economic media.

Volumes published in the series will include both multidisciplinary studies with a broad range of coverage, as well as more focused in-depth analyses of a particular issue in the forest and related sectors. Themes range from globalization processes and international legislation to comparative analyses of regions and countries.

Extreme Conflict and Tropical Forests

Edited by

WIL DE JONG
Center for Integrated Area Studies, Kyoto University, Japan

DEANNA DONOVAN
World Water Assessment Programme, Paris, France

and

KEN-ICHI ABE
Center for Integrated Area Studies, Kyoto University, Japan

Springer

A C.I.P. Catalogue record for this book is available from the Library of Congress.

ISBN 978-94-007-9583-9
ISBN 978-1-4020-5462-4 (eBook)

Published by Springer,
P.O. Box 17, 3300 AA Dordrecht, The Netherlands.

www.springer.com

Printed on acid-free paper

CONTENTS

Tables and Figures vii

Foreword ix

Chapter 1 Tropical Forests and Extreme Conflict 1
Deanna Donovan, Wil de Jong, and Ken-ichi Abe

Chapter 2 Between War and Peace: Violence and Accommodation in the Cambodian Logging Sector 17
Philippe Le Billon and Simon Springer

Chapter 3 Greed or Grievance in West Africa's Forest Wars? 37
Ruben De Koning

Chapter 4 Nicaragua's Frontier: The Bosawas Biosphere Reserve 57
A. Carla Staver, Wil De Jong, and David Kaimowitz

Chapter 5 Past Conflicts and Resource Use in Postwar Lao PDR 75
Yayoi Fujita, Khamla Phanvilay, and Deanna Donovan

Chapter 6 Seeing the Forest for the Trees: Tropical Forests, the State and Violent Conflict in Africa 93
Larry A. Swatuk

Chapter 7 Confronting Conflict Timber 117
Steven Price, Deanna Donovan, and Wil De Jong

Chapter 8 Environmental Damage from Illicit Drug Crops in Colombia 133
María D. Álvarez

Chapter 9 Defoliation During the Vietnam War 149
Goro Nakamura

Chapter 10 Addressing Extreme Conflicts Through Peace Parks 159
Jeffrey A. McNeely

Index 173

TABLES AND FIGURES

TABLES

3.1 Acronyms used in the chapter 39

4.1 Acronyms used in this chapter 64

6.1 Land and forest area for selected Sub-Saharan African countries 98

6.2 Tropical forest area, protected area, and trade in forest products 99

8.1 Estimation of the amounts of solid and liquid chemical reagents used in processing illicit crops during the year 2000 136

8.2 Area affected by illicit crop production and processing, 1998 139

10.1 Examples of negative impacts of war on biodiversity or conservation efforts 162

10.2 Negative and positive impacts of war on biodiversity 165

10.3 Trans-frontier protected areas in tropical forest regions 169

FIGURES

4.1 Nicaragua and the Bosawas Biosphere Reserve 59

5.1 Migration to Vientiane, 1960–1970 76

5.2 Land and forest cover in Phou Phanang 77

5.3 Access to agricultural land and rice production 86

8.1 Illicit crops in Colombia and the world since 1986 135

8.2 Eradication using chemical herbicide in Colombia, by various methods and in other countries 136

8.3 Forests in municipalities where illicit crops have been found 137

8.4 Number of threatened bird species and forests in municipalities where illicit crops have been found 138

FOREWORD

There are many compelling reasons for policymakers to pay more attention to forested regions and invest more resources there. Forests provide valuable products and environmental services and several hundred million extremely poor people live near them. Perhaps the most compelling reason of all, however, is that unless policymakers take forest governance seriously and respond better to the needs of the people living there, these regions will continue to be breeding grounds for violent conflict, banditry, and illicit crops.

From Nicaragua's Atlantic Coast to the jungles of Cambodia, there are several dozen countries around the world that have experienced severe breakdowns in law and order in their forested regions. In many of these cases those breakdowns had widespread economic, social, and political consequences that have threatened entire societies. You would think that after all of the suffering over the last few decades in the forested regions of Colombia, Peru, Mexico, Guatemala, the two Congo's, Liberia, Mozambique, Philippines, Solomon Islands, Nepal, Angola, Rwanda, Nicaragua, Côte d'Ivoire, Myanmar, Bangladesh, India, Sri Lanka, Indonesia, Sierra Leone, Senegal, Sudan, Uganda, and Vietnam people would begin to take note. After all, they don't call it jungle warfare for nothing.

Nonetheless, extreme conflicts in forested regions have just begun to get the attention they deserve. Recent years have seen a flurry of writing and discussion on the topic, and this book edited by Wil de Jong, Deanna Donovan, and Ken-ichi Abe has been able to build on that. The editors managed to pull together a great collection of essays covering a wide variety of the relevant cases and themes, which will undoubtedly be useful for both academics and practitioners.

Some chapters focus more on why forested regions have so much violence, while others look more at the environmental and social impacts of the violence that has already occurred. Various chapters also make clear that the environmental, social, and political effects of armed conflict in forested regions linger on long after the shooting stops. Indeed, post-war periods may well pose the greatest dangers for forest ecosystems. Underlying all this discussion is a constant preoccupation with what might be done to make things better, including everything from sanctions to stop timber exports from funding armed conflict to "peace parks" that encourage dialogue on tense borders.

Those of you worried about conflict should read this book to discover new ways to address it. For those working on forests you can expect to find useful insights into some major factors that are currently influencing their condition. Specialists focusing on specific regions such as the Congo Basin, Mekong Delta, or the Amazon can learn a lot from seeing familiar stories presented in innovative new frameworks. This research has certainly enriched my own thinking about these issues, and I am sure it will enrich your own views as well.

David Kaimowitz

Former Director General, Center for International Forestry Research.
Program Officer Environment and Development, Ford Foundation
Regional Office for Mexico and Central America.

CHAPTER 1

TROPICAL FORESTS AND EXTREME CONFLICT

DEANNA DONOVAN, WIL DE JONG, AND KEN-ICHI ABE

> "If trees could speak, they would cry out that since they are not the cause of war, it is wrong for them to bear its penalties."
>
> —Hugo de Groot, seventeenth century

1 INTRODUCTION

Is it coincidence that of all the conflicts of the twentieth century at least half have been in forested areas? Some people have been asking if this is happenstance or whether there is something about forests that attracts discord. The phenomenon is not only widespread but very worrying, both from the perspective of the threat to the unique ecosystems affected as well as the communities involved. Given these circumstances, a better understanding of the linkages between tropical forests and extreme conflict could assist policymakers and practitioners alike in grappling with the major issues associated with conflict resolution and environmental conservation. Better informed decision making would contribute greatly to addressing security concerns, a key issue on the agenda of many governments and international agencies and organizations around the world today.

Over the last 50 years civil war has become the most common type of conflict. Indeed, the international intervention in Iraq, so prominent in the news of late, is fast becoming a civil war, if not already. There seems to be a tendency for some conflicts initially internal to become internationalized as one or more foreign powers become involved directly or indirectly. Close examination of these incidents reveals that forests and extreme conflict[1] show a strong relationship. Three-quarters of Asian forests, two-thirds of African forests and one-third of Latin American forests have been affected by violent conflict. Currently there are violent conflicts on almost every continent. In Latin America armed violence is prominent in the forest areas of Colombia, in the forest

[1] Extreme conflicts are those that go beyond localized conflict with only local actors. Civil war is an extreme conflict, defined by Singer and Small (1982, cited in de Koning, 2006) as war with at least 1,000 battle related deaths, of which 5% are among rebel and government forces. The cases discussed in this book all fall within this category.

W. de Jong, et al. (eds.), Extreme Conflict and Tropical Forests, 1–15.

foothills of the Peruvian Andes, and in several Central American countries (see Chapter 4). On a smaller scale, but locally significant, are conflicts in the forest regions in Bolivia and Brazil (e.g., Ruiz, 2006). Conflict across Asia including in Afghanistan, Pakistan, India, Nepal, Myanmar, and Indonesia, among other countries, regularly makes the headlines. Almost daily we receive reports of violence in Africa, including in the Sudan, Uganda, the Congo, Liberia, and Nigeria (e.g., see Chapter 3). Over the past two decades armed conflicts have raged in the forested regions of Burundi, Cambodia, the Central African Republic, Guatemala, Mexico, Mozambique, Nicaragua, Peru, Rwanda, the Solomon Islands, and Surinam.

Together the countries affected by violent conflict are home to more than 40% of the world's tropical forest. Indeed, the last remaining tropical forests are located in areas that over the past two decades have been subjected to such conflict. Tropical forests have been not only the venue of vicious confrontation but also the victim. But what is the relationship? Forest resources are exploited by both refugees and combatants alike for provisions and marketable raw materials. Trade in tropical timber has financed violent conflict in the Democratic Republic of the Congo (DRC), Liberia, and Sierra Leone in Africa (see Chapters 3 and 7) and Cambodia and Myanmar in Asia (Le Billon, 2001, Chapter 2), among other countries. Nontimber resources, such as wildlife in Africa and medicinal plants in Nepal, have also been exploited by insurgents while forest in Colombia, Peru, Bolivia, Laos and Myanmar have been felled in order to open land for the cultivation of illicit drug crops used to finance conflict in these regions (see Chapter 8).

Discourse on the linkages between resources, such as timber, and violent conflict has caught the attention of policy makers over this past decade and has begun to influence both foreign and environmental policy. Certainly conflict timber, or timber that is traded to finance armed combat, has been subject of discussion at the highest international levels, inter alia, the United Nations Security Council (UNSC, 2001). Several publications have called attention to this worsening crisis (e.g., Austin & Bruch, 2000; Blom *et al.*, 2000; Price, 2003; Shambaugh *et al.*, 2001; see also Chapter 7). Over the past few years the situation vis-à-vis tropical forests and extreme conflict has failed to improve and, with regard to the level of environmental destruction, may even have worsened.

This book takes an in-depth look at some of these conflicts, past and present, which have so closely involved tropical forests. Authors from a variety of backgrounds including people who have spent many years on the ground in tropical forest regions bring their knowledge and experience to focus on this topic. They discuss the impetus as well as the impacts and implications of extreme violence in tropical forest areas. Some too offer suggestions on how the issue, both the conflict linked to tropical forests and the negative impact this may have on the forest and its biodiversity, might be addressed. Ultimately if the forest suffers, society suffers. We hope this book offers some insight to the relationship between forest and society when violent conflict erupts. If we understand better the role of forest, the reasoning of the combatants, and the result of this connection, perhaps we will be better prepared and more politically motivated to take steps to address these activities and their affect on the tropical forest. In mitigating the impact of extreme conflict in tropical forest, we spare not only the natural environment but the societies that depend upon these resources for their well-being and livelihood.

This first chapter presents an overview of the several themes related to tropical forests and extreme conflict linking them to the various chapters in this book. This chapter summarizes some of the discussion centering on the following two key questions: (1) Given the observed correlation between tropical forest and extreme conflicts, what role do tropical forests have in these conflicts, if any and (2) what is the impact of extreme conflicts on tropical forests, and discusses these questions vis-à-vis the subsequent chapters in the book. Divided into five sections this introductory chapter first explores the historical continuity of the key issues that characterize the tropical forest–extreme conflict link. The next section summarizes various theoretical debates on the issue, introducing evidence from the various chapters that follow. The subsequent section turns to the impact of extreme conflicts on tropical forests while the final section explores the implications of the forest–conflict link.

2 CONFLICTS AND FORESTS IN HISTORY

Most of the chapters in this volume examine contemporary cases to explain the link between violent conflict and tropical forest. This may give the impression that this phenomenon is only of the modern age. In fact forest resources have figured prominently in wars for thousands of years as, among other roles, a source of raw materials, a venue of war, or as a buffer zone separating opposing groups (see Chapter 10). Recognition of the severe environmental impacts of extreme conflict dates back several centuries at least. The history of extreme conflicts and tropical forests is relevant, as it sheds light on the evolution of links and causalities. The theoretical debates that have emerged over the past several years regarding the relationship between natural resources and extreme conflict brings us up-to-date in our review of this relationship.

The significance of the continuum of the forest–extreme conflict link emerges in reviewing how this relationship evolved. In most areas from the perspective of the ruling group the natural resources, including the forest, to which it has access are important sources of wealth and power. Considered a "strategic" resource since earliest times, forest provided impediment to invaders as well as refuge and resources to residents and rulers alike (Donovan, 2003a). From the forest came the raw materials for construction of ramparts and battering rams, the charcoal to produce the metal and subsequently the munitions and the timber and naval stores to build the navies, which played a key role in the defense strategy of the major powers well until the end of the nineteenth century (Perlin, 1991). Invading armies also looked to local resources, including forests, to meet their needs, both for the provisioning as well as the protection of their forces (Christian, 1945; Clausewitz, 1987; Peluso, 1992; Thuy, 2000). For societies under siege local forests could often provide substitute resources to ease the hardship of wartime blockade (Offenberg, 2003; Tsutsui, 2003). Even after cessation of violent conflict, forests have continued to play an important role providing the victors the resources to reward family, friends, loyal followers and other favorites for their support (Donovan,

2003a; Gibbons, 1985). The history of Southeast Asia illustrates these points. Since the earliest days of international trade, forests provided the high-value products that the governing powers could barter or sell for the cash for luxuries, armaments and other goods which could be sold to augment government income.

For many of the governments in Southeast Asia the trade in forest products remained a major source of government revenues for several centuries (Boomgaard, 1995; Donovan, 2003b; Reid, 1993). Attacking and destroying local kingdoms and sultanates European countries initially occupied territories in Asia in the late fifteenth century largely in order to gain control of the forest rich in natural products so valued in international trade at that time, such as spices and incense. Government revenues from these resources, whether in the colonial era, which lasted several hundred years, or postcolonial era since World War II, have been expended in large part on maintaining the existing power structure through support of a strong military presence. Thus the traditional link between the tropical forest and the military has persisted well into the modern era with many modern nations supporting their armies with proceeds directly from natural resource exploitation (e.g., Barber & Talbott, 2003; Nette, 1995; Van Zorge, 1999; see also Chapters 2 and 3).

An issue seldom addressed in the contemporary tropical forest–extreme conflict debate is how the nature of the relationship has changed over time especially due to the impact of technological progress. With the evolution of military hardware and communications technology, the main instruments of warfare have changed and so too the role of the forest. Whereas initially forests provided the raw materials directly needed for conflict, increasingly forests have become important sources of high-value products, such as timber, that are sold to provide the hard currency required for purchasing modern armaments (UNSC, 2001; see also Chapters 2, 3, and 7). While formerly forest products were obtained from remote areas by way of often lengthy and complex supply chains involving several ethnic groups, some traditionally hunter-gatherers and others traders, today improvements in transportation and communications technology have led to a situation in which virtual strangers can arrive in the heart of the forest to harvest valuable forest products with local people often being bypassed. Increasingly local communities resent this intrusion and disenfranchisement and are resisting these outside forces. It can be expected that in the future technologies currently under development will, for instance, allow better monitoring of log extraction and movements (see Chapters 3 and 7) and thus may again change the relationship between extreme conflict and forests in the future.

Most remaining natural forest today exists in remote border areas, distant from central government headquarters. Although timber and charcoal may no longer be used directly for the production of weaponry, the traditional view of the forest as a source of strategic materials still underpins many a government attitude toward forest management. The remote frontier forests, however, are commonly inhabited by indigenous communities, and people driven to the margin, economically as well socially, politically and geographically (de Jong *et al.*, 2006). In areas often marginally suited for agriculture, far from regular markets and lacking access to capital, the inhabitants look to the natural resources, essentially the forest, to provide the means, physical and financial, for their survival and influence. Where more than one group, including central government, seeks control of available resources you have the seeds of potential conflict.

Because forest resources have historically been used both to shelter and to sustain insurgents, in some cases the forest has also been targeted for destruction (e.g., see Chapters 9 and 10). Environmental destruction is a long enduring if lamentable strategy of armies seeking to deprive the opposition of refuge and resources. With the objective being to punish and subdue one's enemy—as opposed to obliterating them—a scorched earth policy was a common tactic of much of early warfare. Thus, although not a new phenomenon, due to new technologies the effectiveness of such forest destruction during extreme conflict has increased tremendously over the years.

Throughout history, however, societies have recognized both the significant social as well as economic impacts of environmental destruction accompanying extreme conflict, and in some cases it has voiced its concern. One of the earliest calls for restraint was recorded by a scribe serving the Greek admiral, Nearkhos, exploring India in the fourth century B.C., who noted that "if there is an internal war among the Indians, it is not lawful for them to touch these land workers, nor even devastate the land itself" (Hughes, 1993, p. 18). Much later in seventeenth century Europe Hugo de Groot, the noted Dutch legalist produced the treatise "On the Law of War and Peace" in which he proclaims "If trees could speak they would cry out that since they are not the cause of war it is wrong for them to bear its penalties." Attempts to limit the impact of war on civilians and the environment have been codified since the 1800s. Thus, laws dating from the mid-nineteenth century forbade the use of certain weapons, restricted the means and methods of warfare to protect property, and established rules for the treatment of civilians and prisoners.[2]

After the Vietnam War we have seen the international community respond to the U.S. government's campaign of herbicide spraying in Southeast Asia (see Chapter 9) and to allegations of U.S. attempts to manipulate weather over Vietnam with the ratification of the *Environmental Modification Convention* (1976) and the *Additional Protocol I to the Geneva Convention* (1977), which prohibit actions that could bring about long-term (or long-lasting), widespread and severe environmental damage (Austin & Bruch, 2003). Not until more than a decade later during the Gulf War in 1991 did the international community and concerned scholars have a chance to examine the adequacy of these measures and other international legal structures (Austin & Bruch, 2000). Environmental lawyers Austin and Bruch (2003) in assessing the progress in the development of norms and institutions to prevent, assess and redress wanton environmental damage resulting from armed conflict note not only a lack of clarification of legal norms, but also of the scientific and economic tools and political will to implement them. A more recent, perhaps more positive step may be the 2003 decision by the International

---

[2]For example, the Declaration Renouncing the Use in Time of War of Certain Explosive Projectiles, St. Petersburg, 29 November/11 December 1898; the Lieber Code of 1863; the Hague Conventions of 1899 and 1907.

Criminal Court to investigate companies dealing in so-called "blood diamonds" (BBC, 2003). This may indicate a greater willingness among the international community to take more interest in the larger issue of the role of natural resources in contributing to the prolongation of extreme conflict.

3 EXTREME CONFLICT AND TROPICAL FORESTS

A debate has developed on the contemporary links between tropical forests and extreme conflicts. Before examining the implications of this debate based on the chapters presented in this book, we will first summarize the broader discussion on resources and extreme conflict.

3.1 Extreme Conflict and Environment Links

Disputes over resources are regularly reported as causes of conflict (Anonymous, 2004; Francis, 2002; Woods, 2003; Yeh, 2000). The linkage, however, between extreme conflict and competition for resources is not as straightforward. In quite a few cases competing stakeholders are able to resolve their differences without resorting to violence, suggesting that there are other circumstances that may explain a natural resource–extreme conflict causal link.

During the 1980s a new field of study termed "environmental security" emerged as scientists sought to explain socio-economic influences on the environment and vice-versa. Scholarly debates developed attempting to explain extreme conflict in relation to natural resources. At a more general level, the discourse on extreme conflicts and natural resources can be divided into essentially three factions:

1. The camp that holds that environmental *scarcity* is the key driver of major conflicts. Key authors in this camp are Kaplan (1994) and Homer-Dixon (1994, 1999).
2. The camp in the political ecology tradition which sees complex causes behind the major conflicts dominated by the competition for *power* between different political factions and the dispossession of resources by others. This *power* camp includes key authors such as Salih (1997), Peluso and Watts (2001).
3. The camp that views major conflicts as the result of a combination of key factors creating an enabling environment, including the availability of finance, poor governance, and a low perceived value of human life. In this *opportunity* camp significant authors are Collier and Hoeffler (2001).

A decade ago Robert Kaplan pronounced that the environment would be *the* security issue of the twenty-first century. He argued that foreign policy would increasingly be shaped by "surging populations, spreading disease, deforestation, soil erosion, water depletion, air pollution and possibly rising sea levels ... [which] will prompt mass migrations and in turn incite group conflicts" (Kaplan, 1994). Homer-Dixon, the second prominent proponent of this environmental scarcity hypothesis posited that conflict over scarce resources would dominate the future. However, he foresees a different pathway to such conflict. Scarcity stimulates technological change and therefore per se would not cause problems. Rather, he argued, increased scarcity would lead to decreased economic activity, causing widespread migration, which in turn would weaken states prompting internal conflicts and violence (Homer-Dixon, 1994).

Others have taken a different route to explain the link between the environmental destruction and extreme conflict. Salih (1997) chose to examine the phenomenon in the larger social context. He argued that "dissent about environmental destruction and externalities of resource use [is actually] . . . about rights and resource control . . . ." Political ecologists argue that proclaiming conflict to be "environmental" in origin obscures the political–economic roots of the clash (Fairhead, 2001; Peluso & Watts, 2001). Squarely opposed to the scarcity camp, they hold that it is erroneous to suggest that resource depletion and the resulting poverty lead to conflict. Rather it is the underlying power imbalance and political structures that result in environmental decline. Government policies, they argue, shape and contribute to resource distribution and depletion and land-use choices.

Labeling Kaplan a "resurgent Malthusian," Peluso and Watts (2001), argue that it is not a matter of scarcity but rather control. More specifically, they propose that the relevant question is not *how much* of a resource there is but *who* has access to its benefits. As they point out, "violence is site specific . . . rooted in local histories and social relations yet connected to . . . material transformations and power relations." Conflict erupts where governments have moved in "their" people and alienated and marginalized the existing population, often the traditional users of the subsequently disputed resources. Such scenarios have been repeated throughout history, around the world, and in every culture. Thus, they contend, extreme conflicts are in most cases the result of a long history of power usurpation linked to material control.

Proponents of the third camp observe that most contemporary conflict is civil or internal. Collier *et al.* (2001) conducted statistical analysis of a large number of cases in an attempt to identify correlations between factors that might be causally linked. Their analysis suggests that *opportunity* for rebellion explains many conflicts, while indicators of grievance alone do not. The availability of finance, as they observed, is one of the key factors positively correlated with opportunity. Primary commodities, such as timber and diamonds, in providing opportunities for exploitation and extortion have proved to be a good source of finance. A climate of poor governance, which permits resource appropriation and extortion, substantially increases the risk of conflict. Moreover, money repatriated by emigrants, an important source of funding for dissident groups, also enhances the opportunity for conflict.

A second important factor that emerged from the Collier *et al.* (2001) modeling exercise was the perceived "opportunity cost" of rebellion. It appears that potential combatants perceive the opportunity cost of conflict to be low, especially where rates of male secondary education enrollment and per capita income are low. In effect this reflects the insurgents' perception of the limited prospects and dismal future available to them and thus by inference a low value placed on human life. While such a "nothing to lose" mentality dominates, it was argued, many individuals in the population are apt to jump at any opportunities for extortion and appropriation of resources. In the final analysis the modelers concluded that "opportunity as an explanation of conflict . . . is consistent with the interpretation of rebellion as greed [and] ... grievance ..." motivated (Collier & Hoeffler, 2001).

3.2 Extreme Conflict and Tropical Forest Links

The many cases of apparent links between tropical forest and extreme conflict warrant their assessment in the light of the environmental security debate. Five chapters in this book are relevant in this context. The statistical correlation between the conflict and tropical forest is debated, with some arguing that cases of civil war are more common in countries *without* tropical forest than in countries *with* tropical forest (Chapter 3). This observation in itself, however, does not exclude a possible causal role for tropical forests vis-à-vis civil war in countries rich in tropical forest. In Chapter 3, de Koning investigates tropical forest conflicts in West Africa. Detailed analysis of these cases suggests that the central argument held by the Collier and Hoeffler camp, needs to be turned around for these cases. Rather than opportunities for greed being the central cause of rebellion, fostering an environment of grievance, it is the condition of grievance, caused by political inequality, lack of opportunities, and the abuse of power by traditional leaders, that provides the root cause of conflict in several of the West African cases. Under such circumstances, political opportunists benefit from such grievance by mobilizing rebel groups largely to further their own political interests. The rebel groups form in forest regions because that is where much of the grievance is located and where the disenfranchised have been pushed. The coincidence of grievance and forests thus is not accidental. Remote forest regions often have limited State presence, as noted by Staver *et al.* (Chapter 4). Where this limited presence coincides with stagnated nation building (Chapters 3 and 6), remote forest regions have two contributing forces that fuel grievance and that may enable this sentiment to contribute to the causes ultimately leading to civil war.

Swatuk, in Chapter 6 argues that trying to find explanation for the tropical forest–extreme conflict link through a narrow focus on the cases themselves misses the critical point that African nation states are a cultural implant that is product of a colonial legacy and a neoliberal vision of the appropriate political and economic world order. These conditions are the root causes of why extreme conflicts emerge in many of the African countries that are plagued by civil war. Some mechanisms in these processes may be linked to natural resources according to de Koning (Chapter 3), because dependency on natural resources, for instance tropical timber, he argues, is a contributing factor to the failed nation building process. Commonly governments that rely on natural resources revenues as opposed to taxes for income are often little worried about issues such as accountability and as a result reduce the delivery of services, contributing to the grievances that eventually lead to civil unrest and in some cases outright war.

In many instances forests play a more prominent role in civil wars as a venue for battle because forest cover benefits groups that pursue guerilla-style warfare. Warring factions also seek to control forest resources in order to obtain the timber or minerals that can be used to finance their operations and living costs. Some of the rebel actions may also be motivated by a desire to capture resources for personal enrichment. De Koning

(2006) suggests that the role of forest as an appropriate venue for warring factions applies to seven of the eight cases in Africa that he has analyzed. Wealth accumulation through the sales of timber, to finance war or for personal profit, applied to three of the eight cases analyzed. In all the cases, however, the underlying causes of the conflicts were not primarily interests in resources. Where forests or political and social conditions in forested regions are not a key causal factor in conflicts, over time forests may become a key element because of the diverse and unique goods and services they offer.

Even though de Koning (2006) finds little evidence of scarcity leading to conflict, and as a consequence rejects the assertions that tropical forest resource scarcity is a direct cause of extreme conflict, two cases discussed in this volume do suggest that scarcity as evidenced by the competition for resources may contribute to extreme conflict under certain circumstances. Staver *et al.* (Chapter 4) show that after the civil war in Nicaragua different groups looking to settle in the North and South Atlantic Autonomous Regions and the Rio San Juan Department came into conflict over the issue of limited forestland. A very similar process took place in Northwest Vientiane Province in Lao PDR (Chapter 5). The actual conflicts discussed in the two chapters are directly caused by resources scarcity. In both cases, however, the recent history of fighting, but also a history of distrust and a breakdown in the rule of law contribute to explaining why these conflicts, especially in Nicaragua, turned violent. In both cases civil wars were the cause of the emerging contestation over limited resources which then became the direct cause of conflict.

4 LEGACIES OF CONFLICTS IN THE FOREST

The extreme conflict–tropical forest debate is important to be able to understand causal links in this relationship, links which could possibly explain causes of conflicts, but also the role that forests play in perpetuating conflicts. Numerous studies have explored the impacts of war on the environment.[3] The plight and impact of refugees from war zones is particularly moving (Biswas & Tortajada-Quiroiz, 1996; Blom & Yamindou, 2001; Sato *et al.*, 2000; Wilkinson, 2002). Longer-lasting impacts especially with regard to the spread of disease, both in human and plant and animal populations, are less well-known but equally as important (Diamond, 1997; Karlen, 1995; Pilcher, 2004). Several of the chapters of this volume speak not only to such impacts but the longer-term implications of extreme conflict in tropical forest regions. Clearly we see that extreme conflict in tropical forests brings effects that have a long and pervasive legacy.

The most salient characteristic of Colombia's recent history is the persistence of armed conflict, albeit of varying intensity. Conflict between belligerent groups of both the right and the left of the political spectrum is not only a key constraint to development in remote areas, but has created a political and economic climate favorable to the

[3] A few recent publications include Blom and Yamindou (2001), Blom *et al.* (2000), Boer *et al.* (2000), Draulans and van Krunkelsven (2002), Frkovic (1999), Milius (2000), Perkins (2004), Price (2003), Saba (2001), Tsutsui (2003).

production and processing of illicit crops, the proceeds of which support the guerilla armies and perpetuate conflict. As Álvarez explains in Chapter 8, damage to the forest though indirect, that is, not the result of combat, is almost more insidious. Large areas of natural forest, some of which is in designated national parks, are felled to provide land for coca production and mountain rivers are polluted with effluent from coca processing and drug production. Although difficult to quantify given the political situation in these areas, the impact of these activities is clearly significant.

Nakamura in Chapter 9 revisits the recent history of Southeast Asia providing a graphic description of the consequences of American herbicide use in this region during the Vietnam War. The impacts of even just a few years of such activity are still with us, several decades later. Focusing on the lingering effects of these chemicals on the environment and the populations involved, he shows how the damages are not limited to the targeted forests, but affect the families of Vietnamese civilians and combatants who lived in or near those areas as well as of the American soldiers involved in the operation. Le Billon (Chapter 2), Price *et al.* (Chapter 7) and Staver *et al.* (Chapter 4) also show that the impacts of war on forests go beyond the actual period of active conflict and also affect forest governance. As Le Billon (Chapter 2) notes, both during and after armed conflict the *instrumentalization* of violence created socio-political conditions in which corruption and forest exploitation thrived. Forest concessions granted in many cases to reward political patronage generate a deep-seated resentment among the population at large. The examples presented by Le Billon (Chapter 2), and to some extent also by Price *et al.* (Chapter 7) and Staver *et al.* (Chapter 4), demonstrate clearly the detrimental impact of extreme conflicts on forest governance.

Describing conditions in Lao PDR in the post-Indochina War era, Fujita *et al.* (Chapter 5) examine how the impact of war on forests does not end with the cessation of violent conflict. With significant influence on land-use patterns and livelihood conditions in all affected areas, the impacts of wartime displacement and postwar resettlement are long-lasting. The impact of the social and economic adjustments required in the postwar period inevitably reverberate across the landscape as forest land is redistributed and the new forest management initiatives are instituted. In summary, all these chapters illustrate how forest governance remains shattered after extreme conflict and is only very slowly restored. During the postwar period valuable forest resources become fair game for remaining insurgents and for entrepreneurs attempting to benefit from the vacuum of governance. Similarly, surviving civilians trying to rebuild their lives after extreme conflict turn to exploitation of forest resources as few alternatives exist in a war-ravaged economy (see Chapters 3 and 4).

McNeely (Chapter 10) highlights the fact that extreme conflicts involving tropical forest occur often on frontiers or in border regions, generally areas of dynamic change and weak government control. Even though wartime damage may be minimal, as logging and poaching are largely restricted due to widespread violence, for the forest the respite is only temporary. In the postconflict era nearly all sectors of society look to the forest resources for the means to rebuild their lives. Thus damage to forest ecosystems and biodiversity often intensifies. The author suggests that the development of "peace parks" could contribute to fostering reconciliation, rebuilding the economy and preserving the environment. The suggestion that extreme conflict actually protects forest, in some instances, should be treated with caution. It may indeed be true that widespread violence, real or threatened, discourages people from entering forests to harvest materials legally or illegally during actual conflict (see Chapter 10). Extended observation, however, reveals that this reprieve is but a brief respite and all too transitory. As several authors, contributing to this volume (see Chapters 2, 4, 5, 7, and 10) and elsewhere (e.g., Price, 2003), have pointed out, predation of forest resources often accelerates after the cessation of armed struggle.

With the end of extreme conflict, governments often try to appease former insurgents and secure political support by granting extraction rights or forest land distribution to supporters. Refugees and other migrants must be accommodated and are often given so-called "abandoned" fields or "unused" forest land to start new lives. Such grants are often at odds with the traditional-use rights of indigenous people in the area, and thus contribute to grievance and set the stage for future conflict. Demobilized militia, often individuals with few skills, less assets and even less schooling, may become involved in illegal logging, wildlife poaching, smuggling or other illicit activities which further threaten already weak government institutions. Blatant corruption and irresponsible management and disposition of forestland and resources may contribute to a sense of grievance among forest-based communities. The low level of technology required to extract most forest resources, and the aggrieved attitude of the marginalized and disenfranchised elements of the population, often lacking a means to produce wealth, influence moves to appropriate forest resources and thus promote conflict. While extreme conflict may be limited, an underlying culture of violence that prevails during a period of ceasefire or negotiated stand-off often creates a climate of fear that makes the conditions in the forest little different than during periods of extreme conflict. Thus, the social, economic and environmental disruption caused by the conflict of regime change, destroy or distort opportunities for development long after violent hostilities subsides.

5 CONCLUSIONS

This introductory chapter and the subsequent chapters in this volume, provide some answers to the two questions formulated in the introduction. The role of tropical forests in forest-related conflicts is complex and changing. Since earliest times forests have been and continue to be used as sanctuary both for civilians evading conflict and combatants eluding capture as well as a source of sustenance for both groups. Under the constant threat of attack from roaming militias, refugees defer crop production and like the militia "live off the land," hunting and gathering what the forest offers. For those able to flee far from conflict zone and to start a new life, distant forest offers the promise of land that can be cleared and converted to crop production with felled and burnt forest vegetation providing the nutrient-rich ash needed to kick-start crop production and begin life anew.

There is little evidence that forest scarcity is a cause for major conflict, as posited by some of the environmental scarcity theorists. In a postconflict setting, however, contestation over limited areas of forestland may be a direct cause of conflicts, which may become violent especially in regions with a recent history of violence and the weakness of postwar governance. Ample evidence exists of the contestation over forest control and access leading to violent conflict. There are historical circumstances that contribute to these conditions. Through the ages the forest has been perceived as part of the national patrimony and a strategic resource. Violence is particularly widespread in forested regions because in general rich forests remain only in areas far from mainstream commercial activity and central government control. These areas, often frontier areas or border regions, are generally populated by ethnic groups marginalized not only geographically but politically and socially as well as economically. In such areas property rights are often poorly developed and differentially defended. Disaffected populations often have only limited allegiance to the central government and look to other, local authorities (warlords) to perform traditional government functions. Competition between rival power brokers and banditry maintains a high level of violence. The ready availability of natural resources—timber, petroleum, diamonds, among others—plus government efforts to promote settlement as a way to lay claim to such areas attracts outsiders. Conflict between long-time inhabitants and newcomers arises as government efforts to develop these areas often ignore local people and their claims to traditional-use rights in forest areas. Thus, conditions of poor governance and a weak institutional environment in many tropical forest areas as much as competition for resources promote conflict. Illegal forestry activities, condoned or uncontested, not only contribute to financing ongoing conflict but to fueling resentment, the seeds of future conflict. Widespread corruption and violence plus the decline of regulatory authority also impede regular, responsible forestry and conservation programs.

Tropical forests tend to be drawn into conflicts, even if they have a limited role as a causal factor in initiating conflict. Given their physical attributes of isolation and inaccessibility, and thus insulated from government control, forests provide excellent cover and therefore their traditional role as a source of shelter may persist. It seems, however, that the current role of forest as the source of financing is the most disturbing and disdainful. The link between forests and power, which is what is at stake in most of the major conflicts, is through forest resources (mainly timber) and commercial interests. Those who seek to gain or maintain power must have the means to impose their will and for many the means or financial resources required for this task will be derived from the liquidation of their nation's natural assets. This process is increasingly facilitated by the advances in technology worldwide and the ready availability of unscrupulous commercial collaborators.

This aspect of the tropical forest–extreme conflict link, provides key elements to the answer of our second question: What is the impact of extreme conflicts on tropical forests? The nature of this impact has changed during history and is changing still. It is not improbable to predict that wide-scale defoliation, as was carried out less then four decades ago during the Vietnam War, will not be repeated, because it will be considered unacceptable worldwide. The exploitation of forest resources to finance armed conflicts or to feed people, because forest governance breaks down or because frightened refugees must flee armed conflicts elsewhere, will probably continue to be key factors contributing to the impact of conflict on tropical forests. We do, however, have an opportunity to reverse or at least reduce some of the negative impacts. Addressing the issue of resources exploitation for financing violence, the international community needs to stand firm and united behind boycotts and other legal actions to deny belligerents the means to escalate conflict. Some signs are hopeful. Charles Taylor was taken into custody and put on trial in the Netherlands. A Dutch timber merchant, Gus van Kouwenhoven, has been found guilty of breaking the UN arms sanction on Liberia and sentenced to 8 years in jail for selling weapons to Charles Taylor in exchange for logging concessions.

But there remains a need to support all feasible initiatives for conciliation, compromise and conservation, such as the peace parks, and to better anticipate and manage the impact of refugees on the environment. The chapters in this volume highlight the need to address forest governance and management issues during any early negotiations to end violent conflict. Subsequent international intervention, such as peacekeepers, must be made aware of environmental issues and must coordinate with conservation agencies and local groups to prevent postconflict depredation of resources, to secure local peoples traditional rights, and to establish viable institutions of local governance that can safeguard forest resources to ensure rural people's future well-being and development. Preventing conflict and reducing its duration and impacts are among the critical challenges today facing advocates and practitioners of sustainable development and environmental conservation.

REFERENCES

Anonymous. (2004, Aug 28). Darfur rebels: No angels. *The Economist,* p. 34.

Austin, J. E., & Bruch, C. E. (Eds.) (2000). *The Environmental Consequence of War: Legal, Economic and Scientific Perspectives.* Cambridge, UK: Cambridge University Press.

Austin, J. E., & Bruch, C. E. (2003). Legal mechanisms for addressing wartime damage to tropical forests. In S. V. Price (Ed.), *War and Tropical Forests: Conservation in Areas of Armed Conflict* (pp. 167–199). Binghamtom, NY: Food Products Press.

Barber, C. V., & Talbott, K. (2003). The chainsaw and the gun: The role of the military in deforesting Indonesia. In S. V. Price (Ed.), *War and Tropical Forests: Conservation in Areas of Armed Conflict* (pp. 137–166). Binghamtom, NY: Food Products Press.

BBC. (2003). *Firms Face Blood Diamond Probe.* Accessed from news.bbc.co.uk/go/pr/fr/-/2/hi/business 3133108.stm.

Biswas, A. K., & Torajada-Quirioz, H. C. (1996). Environmental impacts of the Rwandan refugees on Zaire. *Ambio, 25*, 403–408.

Blom, A., & Yamindou, J. (2001). *A Brief History of Armed Conflict and Its Impact on Biodiversity in the Central African Republic.* Washington, DC: Biodiversity Support Program, USAID/WWF.

Blom, E., Bergsmans, W., Dankleman, I., Verweij, P., Voeten, M., & Wit, P. (Eds.) (2000). *Nature in War: Biodiversity Conservation during Conflicts.* Mededelingen no. 37. Amsterdam: Netherlands Committee for International Nature Protection.

Boer, W. F. de, Ntumi, C. P., Correia, A., & Mafuca, J. M. (2000). Diet and distribution of elephant in the Maputo Elephant Reserve, Mozambique. *African Journal of Ecology, 38*(3), 188–201.

Boomgaard, P. (1995). *The VOC Trade in Asian Forest Products in the 17th and 18th Centuries*. Leiden: Koninklijk Institute voor Taal, Land en Volkenkunde.

Christian, J. L. (1945). *Burma and the Japanese Invader*. Bombay: Thacker & Company.

Clausewitz, C. von. (1987). *On War*. Harmondsworth: Penguin Books (originally published 1832; edited with an introduction by Arnold Rappaport).

Collier, P., & Hoeffler, A. (2001). *Greed and Grievance in Civil War*. Washington, DC: World Bank.

Collier, P., Hoeffler, A., & Söderbom, M. (2001). *On the Duration of Civil War*. Policy Research Working Paper 2681. Washington, DC: World Bank.

de Jong, W., Tuck-Po, L., & Abe, K.-i. (Eds.) (2006). *The Social Ecology of Tropical Forest Migration*. Kyoto: Kyoto University Press and Trans Pacific Press.

de Koning, R. (2006). *Troubled Woods: What does the forest have to do with conflict in Western and Central Africa* unpublished manuscript.

Diamond, J. (1997). *Guns, Germs and Steel*. New York: N.W. Norton.

Donovan, D. G. (2003a). Evolving role of forests at the edge of empire: The case of Nepal. Paper presented at the *International Conference on Forest and Environmental History of the British Empire and Commonwealth,* Institute of Development Studies, University of Sussex, Mar 19–21, 2003.

Donovan, D. G. (2003b). Trading in the forest: Lessons from Lao history. In L. Tuck-Po, W. De Jong, & K.-i. Abe (Eds.), *Political Ecology of Tropical Forests of Southeast Asia: Historical Perspectives* (pp. 72–106). Kyoto: Kyoto University Press and Trans Pacific Press.

Draulans, D., & van Krunkelsven, E. (2002). The impact of war on forest areas in the Democratic Republic of the Congo. *Oryx, 36*(1), 35–40.

Fairhead, J. (2001). International dimensions of conflict over natural and environmental resources. In N. L. Peluso & M. Watts (Eds.), *Violent Environments* (pp. 213–236). Ithaca: Cornell University Press.

Francis, D. R. (2002, Dec 5). Fueling war. *Christian Science Monitor*, 11.

Frkovic, A. (1999). The mortality of the European brown bear (*Ursus arctos* L.) in mountainous regions of Croatia during (1991–1995) and after (1996) the Civil War. *Sumaiski-List, 123*(11/12), 565–572.

Gibbons, E. (1985). *The Decline and Fall of the Roman Empire*. Harmondsworth: Penguin Books.

Homer-Dixon, T. (1994). Environmental scarcity and violent conflict: Evidence from cases. *International Security, 19*, 5–40.

Homer-Dixon, T. (1999). *Environment, Scarcity and Violence*. Princeton: Princeton University Press.

Hughes, J. D. (1993). Early ecological knowledge of India. In A. S. Rawal (Ed.), *Indian Forestry: A Perspective* (pp. 13–28). New Delhi: Indus Publishing.

Kaplan, R. D. (1994, Feb). The coming anarchy: How scarcity, crime and overpopulation and disease are rapidly destroying the social fabric of our planet. *Atlantic Monthly,* 44–76.

Karlen, A. (1995). *Man and Microbes*. New York: G.P. Putnam.

Le Billon, P. (2001). The political ecology of war: Natural resources and armed conflicts. *Political Geography, 20*, 561–584.

Milius, S. (2000). When biologists get bombed. *Science News, 158*(5), 78–79.

Nette, A. (1995, Dec 15). Development—Laos: Military Inc. Flexes Economic Muscle. World News-InterPress Service, from www.oneworld.org.

Offenberg, K. (2003). Holstankstellen in den 1930er und 1940er Jahren [Wood Service Stations in the 1930s and 1940s]. *Forst-und-Holz, 58*(23/24), 717–719.

Peluso, N. (1992). *Rich Forests, Poor People: Resource Control and Resistance in Java*. Berkeley: University of California Press.

Peluso, N., & Watts, M. (2001). *Violent Environment*. Ithaca: Cornell University Press.

Perkins, S. (2004). Thin soil: Desert's fragile crust takes millennia to form but only moments to destroy. *Science News, 165*(1), 11–13.

Perlin, J. (1991). *A Forest Journey: The Role of Wood in the Development of Civilization*. London: Harvard University Press.

Pilcher, H. (2004, Apr 6). US army may have killed Italian trees: Deadly fungus hitched TransAtlantic lift with American troops. *Nature—Science Update*, from www.nature.com/nsu/040405/040405-1.html.

Price, S. V. (Ed.) (2003). *War and tropical forests: Conservation in areas of armed conflict*. Binghamtom, NY: Food Products Press (co-published simultaneously as the *Journal of Sustainable Forestry, 16*(3/4)).

Reid, A. (1993). *Southeast Asia in the Age of Commerce 1450–1680. Vol. 2. Expansion and Crisis*. New Haven: Yale University Press.

Ruiz, S. (2006). *Rentismo y Conflictos en el Norte Amazónico Boliviano*. Santa Cruz, Bolivia: Center for International Forestry Research.

Saba, D. S. (2001). Afghanistan: Environmental degradation in a fragile ecological setting. *International Journal of Sustainable Development and World Ecology, 8*(4), 279–289.

Salih, M. (1997, May). Politics of poverty management and environment: Displacement by conservation. Paper presented at the *Workshop on Politics of Poverty and Environmental Interventions,* Nordic Africa Institute, Institute of Social Studies, The Hague, Netherlands.

Sato, H., Yasui, K., & Byamana, K. (2000). Follow-up survey of environmental impacts of the Rwandan refugees on Eastern DR Congo. *Ambio, 29*(2), 122–123.

Shambaugh, J., Oglethorpe, J., & Ham, R. (with contributions from Sylvia Tognetti) (2001). *The Trampled Grass: Mitigating the Impacts of Armed Conflict on the Environment*. Washington, DC: Biodiversity Support Program, USAID/WWF.

Thuy, T. T. (2000). Rice cultivating and cattle raising in Tonkin in the first half of the 20th century. In J. P. Bassin, J. D. Goacometti, K. Odaka & S. R. Clark (Eds.), *Quantitative Economic History of Vietnam, 1900–1940*. Tokyo: Institute of Economic Research, Hitosubashi University.

Tsutsui, W. M. (2003). Landscapes in the Dark Valley toward an environmental history of wartime Japan. *Environmental History, 8*(2), 294–311.

United Nations Security Council (UNSC). (2001, April 12). Report of the Panel of Experts on the Illegal Exploitation of Natural Resources and Other Forms of Wealth of the Democratic Republic of the Congo, S/2001/357.

Van Zorge, H., & Associates. (1999, Feb 11). Wiranto's army: Marching into unknown territory. *Van Zorge Report*, 4–11.

Wilkinson, R. (2002). Living on the edge. *Refugees, 2*(127), 4–13.

Woods, D. (2003). The tragedy of the cocoa pod: Rent seeking land and ethnic conflict in Ivory Coast. *Journal of Modern African Studies, 41*(4), 641–655.

Yeh, E. T. (2000, Mar). Forest claims, conflicts and commodification: The political ecology of Tibetan mushroom harvesting villages in Yunnan Province, China. *The China Quarterly* (London), *161*, 264–278.

CHAPTER 2

BETWEEN WAR AND PEACE: VIOLENCE AND ACCOMMODATION IN THE CAMBODIAN LOGGING SECTOR

PHILIPPE LE BILLON AND SIMON SPRINGER

1 INTRODUCTION

From the deadly raids launched upon Burundi's capital by Hutu rebels hiding in the nearby Tenga forest, to the multimillion dollar exploitation of teak along the Thai-Burmese border by the Karen National Union, insurgents have repeatedly used forests as a refuge or a source of finance. Located near roads and towns that are of military importance, or along border areas offering political sanctuary, forests provide some of the safest terrain from which to prepare or launch guerrilla operations. Forest products are among the most conspicuous resources financing wars in a post-Cold War era where business and predation have replaced foreign state support (Le Billon, 2001). As rebels take advantage of their location in forests to control or establish logging operations, underfunded or financially self-interested government military forces deployed for counterinsurgency purposes frequently join in. Logging companies also seem to accept a higher degree of risk than entrepreneurs in most other sectors, for instance to access increasingly rare and valuable old growth forests.

The war in Cambodia during the early 1990s became emblematic of links between forests, armed groups and logging companies, due to the pioneering advocacy work of Global Witness (1995a, 1996). This London-based human rights organization carefully collected and widely publicized evidence on the role of the logging sector in financing the war, demonstrating the complicity of neighboring countries, the collusion of Cambodian political parties in power, and the massive costs for the population and the environment. Forests not only offered the main stronghold and source of finance of the Khmer Rouge guerrilla force, before it was militarily ousted in 1979 by Vietnamese troops. Forests, and timber in particular, were the subject of numerous conflicts and accommodations between divergent actors, including poor farmers, enterprising generals, and large regional companies.

Ironically, the end of the century importance of the Cambodia forest sector was itself in part the result of previous hostilities. Despite massive bombing in the eastern half of the country by the United States (US), 20 years of tragic history had arguably saved the

W. de Jong, et al. (eds.), Extreme Conflict and Tropical Forests, 17–36.

forests of Cambodia from the onslaught that fuelled the economic growth of many of its neighboring countries. The lengthy transition to peace during the 1990s also brought about a greater role for forests in the conflict. Although revenues from timber provided the "sinews of war," an "all out" war would have likely impeded logging. Logging interests, while far from the only factor involved, contributed to the prolongation of a "neither war nor peace" stalemate, mutually beneficial to belligerents. The conditions of war gave specific actors the opportunity and impunity to conduct logging operations.

This chapter explores how forests contributed to the prolongation of conflict and to the difficulties of transition to peace in Cambodia, including the financing of the Khmer Rouge, the reconfiguration of politico-economic networks of power, and the causes behind the apparent failure of both the government and the international community to transform this valuable natural resource into a positive factor for peace and reconstruction.

2 RESOURCES AND "SHADOW STATE" POLITICS IN TIMES OF TRANSITION

Countries facing the challenges of a transition from war to peace in the turbulent early 1990s faced, among others, a dilemma for natural resources extraction. Natural resources were generally perceived to be the springboard of economic reconstruction while vast social needs argued for the immediate wholesale liquidation of these assets. Natural resource exploitation, however, remained a contested solution given the growing importance of environmental and sustainable development issues in the global political agenda since the late 1980s. Questions arose over which type of resource exploitation, if any, was preferable; what development policies were best suited; and who should benefit?

As a result of this dilemma the commodification of nature has been placed under critical scrutiny for its effect on the environment and populations. For some, mostly within international development agencies, the commodification of nature is to be rationalized, through a program of management that emphasizes "national" prosperity. For others, mostly nongovernmental organizations (NGOs), grassroots social movements, and nature protection groups, this recommodification of nature is to be resisted, as it is destructive of the customary relationship between local communities and their environment, and results in capital accumulation and a concentration of power. The end result of these conflicting perspectives is generally a diversity of "public transcripts" that are then inscribed through policy documents, media reports, or advocacy movements.

The redistribution of costs and benefits resulting from the commodification of nature is shaped by, among others, a "social polity" that organizes relations of power and resistance. The contestation of this redistribution is more likely during political transition as individual actors jockey to improve their position in a highly unstable social environment, often characterized by the breakdown of law and order, institutions, and even customary rules of social behavior. The commodification of nature is thus shaped by a heavily politicized context that is unlikely to conform to usual rational economic or managerial principles. Rather, it is likely to reflect the chaotic nature of the transition associated with the dissolution of mainstream societal driving forces as influenced by the survival strategies of individual actors. These strategies are in turn informed by a "hidden transcript" that is inscribed into actual practice.[1]

In such context, representatives of the state are likely to adopt a two-faced strategy in which official policies conform to the view of the current "public transcript," while totally different, unofficial practices secure individual or group access to, and control of, the commodification process along the lines of the "hidden transcript." In the realm of natural resources, the state itself can be understood to be conducting both formal governance and "shadow" governance. The formal state can be defined as the official institutional structures and the political and legal system regulating them. The "shadow" state, a term coined by Reno (1995, p. 3) in his study of corruption and warlordism in Sierra Leone, refers to the system through which rulers are "drawing authority from their abilities to control markets and their material rewards." There is no neat division between the two, as actors and relations overlap, but a conceptual division might help to understand the supposedly irrational behavior of the state during periods of crisis or transitions.

This vision of the state corroborates the notion of a state in crisis developed in the 1990s around ideas such as "failed" or "predatory" state (Bilgin & Morton, 2002). These concepts describe the state in terms of its failure to provide public services (e.g., sovereignty, order, justice and welfare), emphasizing instead the self-interest of actors within the state apparatus. The failure of states is attributed to the withdrawal of the Cold War order, the rise of ethno-nationalism, and fiscal constraints such as structural adjustments. An alternative view is to see the apparent failure of the state as the result of shadow state strategies used as coping mechanisms by a political leadership struggling to assert its control over key resources in order to translate its position into effective power, if not comprehensive statehood. From this perspective it is possible to see the shadow state as an answer to the political challenges posed by the neoliberal perspective of government by the market, in which "the market is viewed not merely as a means of allocating goods and services but as a form of social regulation" (Graf, 1995, p. 141). Rather than opposing such dominating paradigms as democracy and neoliberalism, a state polity placed into a crisis or transition phase has an interest in co-opting these in order to benefit from their many opportunities, and in attempting to reshape them into instruments of power. In this regard, Bates (1981, p. 129) has argued that interventions by the state in the political economy of rural Africa served not only as simple political distortions of market forces, but as "political resources . . . which can be used to organize political support." The case of logging in Cambodia illustrates this process, as well as the limits and specificities brought by the context and instrumentalization of war.

3 LOGGING AND WAR IN CAMBODIA

Exploited in some easily accessible areas under the French protectorate and during the early postcolonial era, Cambodian forests still covered nearly two-thirds of the

[1] On "public" and "hidden transcripts," see Scott (1990).

country by the late 1980s. By and large, Cambodia's forests had also survived both the heavy bombing of the US military during the early 1970s, and the Khmer Rouge's subsequent murderous drive for agrarian utopia between 1975 and the end of 1978. Forest exploitation during the 1980s, mostly by Vietnamese occupying forces, remained limited as the country endured an embargo on development aid and international trade imposed by the West. By the late 1980s, at the onset of a new era made possible by the end of the Cold War, forests accounted as a key economic and environmental asset for Cambodia, a country listed as among the poorest in the world.[2] Since then, much of these forests have been consumed by intensive commercial logging as the country reintegrated into the global economy.[3]

The exploitation of Cambodian forests for their timber over the last 15 years shares some similarities with the experience of other Southeast Asian countries. The process of forest commodification was politicized, sustaining the power of the army and the political elite (Broad, 1995; Broad & Cavanagh, 1993; Bryant, 1997; Ghee & Valencia, 1990; Hurst, 1990; Vitug, 1993). Forest dwellers were adversely affected and progressively disempowered by large-scale commercial logging that responded to foreign demand; from Japan in particular (Colchester, 1993; Dauvergne, 1997; Potter, 1993). Cambodia's recent forest exploitation, however, was complicated by its lengthy and difficult transition from war to peace. From the signing of the Paris Peace Agreement between opposing Cambodian factions in 1991 until the end of the first mandate of the newly elected government in 1998, at least USD 2.5 billion worth of timber was exported from Cambodia, roughly equivalent to the nation's average annual gross domestic product (GDP) for that period. By the mid-1990s timber earnings represented an estimated 43% of Cambodian export revenues, more than any other country at that time (FAO, 1997). While such figures might signal an improvement of the economic well-being of the population, a large part of this financial windfall did not contribute to the official process of reconstruction and development as it escaped public taxation, fuelled conflict, broadened wealth disparities, and deepened an environmental crisis (Global Witness, 1995a; Talbott, 1998).[4] Logging, and its environmental and economic consequences, captured the attention of the international media, NGOs, and development agencies. At an international donor meeting the International Monetary Fund (IMF) argued that, "[t]he depletion of the country's most valuable resource [forests] . . . with the direct involvement of the highest levels of government . . . remains the single most critical issue in Cambodia" (Neiss, 1997). As demonstrated by the coup d'etat that shortly followed this statement, logging was not the single most critical issue in Cambodia. Rather, logging was part and parcel of a broader problem of conflict and accommodation within a deeply divided government.

[2] Commercial forests (evergreen and mixed forests) decreased by less than 10% between 1960 and 1990 (FAO, 1994).

[3] Cambodia joined ASEAN on April 30, 1999.

[4] The total public revenue collected between 1991 and 1998 represented only USD 97 million, about 12% of what the government should have received under a proper system of taxation (Ministry of Economy and Finance, 1997; World Bank *et al.*, 1996).

3.1 Setting of a Tragic History

Located between Vietnam and Thailand, Cambodia's physiography consists of a central flood plain that is drained by the Mekong River. The plain is surrounded by low altitude mountains and undulating plateaus covered by evergreen or deciduous forests. A short coastline offers access to the Gulf of Thailand. Cambodia's population, estimated at 12 millions in the late 1990s, grows at an annual rate of 2.8%. Nearly half of this population is under the age of 15 and 80% live in the countryside (Ministry of Environment, 1994). Although not densely populated (62 inhabitants/km^2), the population distribution is extremely unequal with about 85% living in the central plains, leaving vast expanses of sparsely populated forests on the periphery (Ministry of Environment, 1994). Natural resources are central in the livelihoods of the vast majority of the population as economic diversification has only recently begun, due in part to nearly three decades of civil war and international isolation. In the mid-1990s, nearly 50% of the GDP still came from the agricultural sector (Ministry of Economy and Finance, 1997). The livelihood of the average Cambodian still largely depends on rice production and fisheries, and the rhythm of life is attuned to the shift of dry and rainy seasons and the movements of water in the flood plains.

Over the past millennium, Cambodia has been characterized by its rise and decline as a powerful political entity in the region, and over the last 30 years by warfare and genocide. The period as a French protectorate ended relatively peacefully in the early 1950s when power was handed back to the indigenous monarchy. Since then the country attempted in vain to remain outside of the war afflicting the rest of Indochina. Conflict finally erupted in 1969 and ended with the victory of communist forces, known as the Khmer Rouge. Vietnamese forces ousted the Khmer Rouge from power in 1979, and installed and supported a friendly government, the People's Republic of Kampuchea (later renamed State of Cambodia, SOC) under the single rule of the Cambodian People's Party (CPP).[5] The war continued as both China and the West supported the Coalition Government of Democratic Kampuchea (CGDK). The CGDK included remnants of the Khmer Rouge, the pro-Royalist party FUNCINPEC (French acronym for the National United Front for an Independent, Neutral, Peaceful, and Co-operative Cambodia), and the republican Khmer People's National Liberation Front (KPLNF). A political stalemate between the two blocks was resolved in 1991 by a peace agreement enforced through a USD 2 billion United Nations (UN) operation charged with organizing elections that FUNCINPEC ultimately won. Under pressure from the CPP, a coalition government was formed that maintained a military opposed by the Khmer Rouge. After 4 years of cohabitation, a coup d'état returned power to the CPP headed by Hun Sen, a situation that was sustained by the results of elections held in 1998.

With the political end of the civil war in 1991, many international development agencies predicted a new beginning towards "prosperity" for Cambodia, particularly because

[5] For a comprehensive picture of Cambodia's history, see Chandler (1996). On the third Indochina war, see Regaud (1992); for analyses of the UN period see Findlay (1995) and Heder and Ledgerwood (1996).

of the dynamic regional environment and the prospective of joining the economic success of neighboring countries (World Bank, 1992). Much hope was initially placed on the exploitation of natural resources for promoting economic growth and reconstruction. Over the last decade, however, the pattern of exploitation of natural resources, and specifically of timber, has been considered a major failure in the reconstruction process and even an impediment to peace-building due to its association with civil war, endemic corruption, weak tax revenues, severe environmental degradation, human rights abuses, poverty, and food insecurity. Observers recognize this situation as a threat to the development and long-term stability of the country, and to the welfare of its population. Such concerns stem from the negative effects that the mismanagement of natural resources can have not only on the livelihood of the population, but also on the legitimacy and capacity of the state (Global Witness, 1995a; Neiss, 1997; Talbott, 1998).

Capital accumulation through natural resource exploitation is, nevertheless, generally considered necessary to provide the state with adequate means, to carry out reconstruction in countries affected by decades of conflict. For Cambodia, most reports by international development agencies stressed that the country was "well endowed with natural resources" (FAO, 1994, p. ix) that provided great potential for contributing to development and economic growth. Such capital accumulation often faces criticism, however, as it tends to increase the unequal distribution of wealth among the population and negatively affect the environment. Indeed, while the richness of the forests was often stressed, few development agencies after 1989 openly argued for the development of forestry. In stead most took a cautious approach calling for further studies and master plans, if not for outright afforestation (ADB, 1991; UNDP, 1989; World Bank, 1992). The importance given by political leaders, the media, and the international community to the problem of *protecting* forests for improving the transition of Cambodia from war to peace and from poverty to prosperity prompted a significant number of declarations, campaigns, and projects.

This short overview of history provides a rather standard, coarse reading of Cambodia's recent history: a bitter struggle for power that ended thanks to the benevolence of the international community and the resounding success of the UN, which was unfortunately tarnished by the misconduct of the Khmer Rouge and later of the CPP. The limitations of the influence of the international community are nevertheless illuminated with a more detailed analysis (see Brown & Timberman, 1998). Another reading, that of politicians, military leaders, and businesspeople coming to terms with each other to further their personal economic and political interests in this rapidly changing context, could provide a different perspective. In this regard, the dynamics of the forestry sector, one of the most profitable of the Cambodian economy, might provide some insights.

3.2 Evolution of the Cambodia Logging Sector from War to Peace

The commodification of forests in Cambodia came to play different roles over the last quarter century. In the 1980s, Vietnamese forces and companies, domestic demand, and petty trading along the border dominated forest exploitation. While forests have long provided a refuge for insurgents, the situation evolved as forests became more valuable thanks to the growing demand for tropical timber. From the early 1980s, Thai military personnel entered into logging deals with armed factions located along Cambodia's western border in exchange for arms and the protection of refugees. On the other hand, the presence of Vietnamese troops along Cambodia's eastern border greatly confined such activities. Vietnamese troops and companies exploited timber inside Cambodia to export it back home. In between, small Cambodian companies officially organized as cooperatives occasionally struck deals with either the Thai or the Vietnamese to access timber. These dealings, however, remained of minor importance to the different warring factions as they continued to receive the financial backing from their foreign sponsors, such as China in the case of the Khmer Rouge.

In 1988, 3 years before the withdrawal of major external sponsorship, the Khmer Rouge leader Pol Pot stressed that the long-held policy of autarky was no longer suitable for Cambodia:

> Our state does not currently have sufficient capital either to expand its strength or to enlarge the army. We are spending many tens of millions of Baht to augment the assistance of our friends, but that is still not enough and there are many shortages. It is thus imperative that we find ways to develop the natural resources that exist in our liberated and semi-liberated zones as assets to be utilized in the fight against the Vietnamese aggressor enemy. . . . [we need to build] gateways for entry and exit to the outside world . . . roads for external contacts . . . . The resources in each and every one of these areas absolutely must be utilized as assets.[6]

The departure of Vietnamese forces in September 1989 tipped the local balance of power in favor of the Khmer Rouge and other factions in peripheral areas close to the Thai border. A logging ban was declared in Thailand in the same year following a devastating landslide. This conjunction of military opportunity and economic demand set the scene for a sharp increase in logging along the Thai border.

Armed groups opposing the government in Phnom Penh took advantage of this opportunity to counter a *rapprochement* between the Thai and Cambodian Prime Ministers that was threatening their position and future interests. While unsuccessful in toppling the government in Phnom Penh, these military offensives yielded greater territorial control that allowed the Khmer Rouge and other groups to offer forest exploitation permits to primarily private Thai companies, with Thai military units taking a protection fee on the way and high-ranking officials in Bangkok providing a tacit and often financially interested green light.[7] The timber supply from Cambodia became even more essential to Thai forestry interests and Thai imports of Cambodian logs increased a five-fold between 1990 and 1994 (Hirsch, 1995; Royal Forest Department, 1996). Despite progress towards a resolution of the Cambodian conflict on the political

---

[6] Cited in Thayer (1991, pp. 29–30).

[7] For example, the Thai Army Unit 838, in charge of covert assistance to the Khmer Rouge, was reportedly getting USD 1.6 per m^3 of timber imported into Thailand, or about USD 1 million per year (Morello & Thayer 1992, p. 12).

front, fighting on the ground escalated partially to increase territorial control and thus financial revenue.

While agricultural schemes and workshops were also set up in territories "liberated" by the armed groups, the exploitation of natural resources was by far the most profitable activity. All groups attempted to extend their territorial control, and to lease logging concessions to foreign companies, including American and Japanese, but primarily Thai. Timber represented by far the main source of income, which also included gems mining, and the trade of stone carvings from ancient temples, latex, gravel, rice, corn, fish, scraps metal, and nontimber forest products such as sandalwood oil (Le Billon, 1999). When the political and financial backing of foreign sponsors dried up in the early 1990s timber and gems, along with the collaboration of Thai companies, became crucial to the financial standing of factions along the border.

The need for revenues did not end with the signing of a peace agreement in 1991. On the contrary, the mounting insecurity associated with the perspective of political change reinforced a pattern of logging-on-the-run and the sale of massive forest concessions inside the country. Each group attempted to secure logging deals before the elections scheduled in mid-1993. Regional logging companies, from Thailand, Vietnam, Indonesia, Malaysia, Singapore, and Taiwan started to operate in Cambodia, acting as a primer for capitalism in the country. Thousands of Cambodians engaged in logging as the economy became increasingly cash based caused by access to international markets and growing demand, as well as a breakdown in law enforcement resulting from competing, corrupt authorities. Among the various logging groups, the military were often best placed to benefit, due to their location and easy access to guns, manpower, and trucks. Logging and basic sawmill equipment was imported *en masse* from the closing Thai forest industry. Some timber deals reached across both political boundaries and associated competing factions. This announced the mutual accommodation of elites that would dominate the political economy of Cambodia and that of the coalition government in particular, following the UN elections (Ashley, 1998; Hughes, 2003). Soon after the beginning of the peace process, the Khmer Rouge withdrew and resumed fighting. Several initiatives were taken to bring them back into the plan for political settlement. The UN Security Council warned the Khmer Rouge that "no reconstruction aid would go to parties not co-operating in the peace process" (UNSC Resolution 766). This economic threat had no impact, in part, because the Khmer Rouge was awash in cash from its business dealings. As diplomatic initiatives also failed and the UN was unwilling to use force, the international community promoted the use of economic sanctions against the Khmer Rouge.

In September 1992, the transitional government in Cambodia—the Supreme National Council (SNC) composed of the UN and Cambodian factions—decided to pressure the Khmer Rouge directly by placing a moratorium on the export of logs from Cambodia, and bringing exports of sawn timber under the control and monitoring of the United Nations Transitional Authority in Cambodia (UNTAC). The protection of Cambodia's forests was invoked, and the ban was applied to all factions in order not to prejudice a "neutral [political] environment conducive to free and fair elections" (SNC, 1992). The Khmer Rouge rejected the SNC decision on the grounds that it "would in practice constitute a form and measure to penalize the [Khmer Rouge] while leaving the door open for the [CPP] and Vietnam to continue logging as before." Both the KPLNF and FUNCINPEC supported the ban, and the CPP ensured that sawn timber exports were not banned as these financed essential imports. The ban on logging exports was subsequently supported by the United Nations Security Council (UNSC), which also requested the SNC "to consider the adoption of a similar moratorium on the export of minerals and gems in order to protect Cambodia's natural resources," and asked Thailand to prevent the supply of petroleum products to the Khmer Rouge (UNSC Resolution 792). The SNC later decided to place a moratorium on the extraction and export of mineral resources, although this was seen as virtually impossible to implement. The gem-mining area was, however, very limited in size, and preventing the access of large mechanized equipment would have drastically reduced extraction rates. An attempt in 1993 to ban the international trade in Khmer Rouge gems by the Cambodia Study Group, a US-based NGO, failed for lack of political support in the US and the European Union (EU). The proposed ban aimed at influencing the behavior of the Thai National Security Council towards the Khmer Rouge by undermining its business interests with the Cambodian insurgents and legitimizing those with the Royal Government of Cambodia (RGC).

Thai political and economic interests affected by the ban, including the Thai Minister of the Interior, General Chavalit Yongchaiyut, lobbied for a postponement of these measures. Threats of job and investment losses by Thai interests were backed by documents that exposed parts of a complex web of relationships between Thai business, the Thai military, and Cambodian political parties. These documents also revealed that many Thai companies dealt with the CPP for some of their operations within Khmer Rouge and FUNCINPEC territories. Ultimately, the Thai government declared that it would abide by the UNSC resolution, but refused to allow UN monitors onto its soil and reaffirmed that "the door should not be closed to negotiation" with the Khmer Rouge. In light of economic sanctions against the Khmer Rouge, the Thai Minister of Foreign Affairs had declared earlier, that "a frustrated UN is looking for a scapegoat in Thailand" and that the Thai military resented "the way UNTAC went about its business [in controlling Thai activities in Khmer Rouge areas]."[8]

Effective backing of UN sanctions on the Thai side was unlikely. Thailand's history of competition with Vietnam over Cambodia has meant that the Thai authorities have traditionally lent their support to Cambodian parties with an anti-Vietnamese stance. Thai companies heavily invested in, and handsomely profiting from Cambodian ventures, along with prominent government and military personalities, became a powerful lobby against the strict implementation of sanctions. Furthermore, the balance of power between the civilian and military authorities in Thailand remained precarious, and the leverage of politicians in Bangkok over military units in border areas and over

[8] "The Thai military had clumsily attempted to pressure the UN to delay implementation of the ban by forcing UN supply flights to give three-day notice before entering Thai airspace [starting on 1 December 1992]" (cited in Findlay 1995, pp. 43–44). The Thai government cancelled the measure after 4 days.

businesses engaged in cross-border trade was weak. Finally, the Thai authorities also supposedly feared the prospect of a cornered Khmer Rouge, which might seek revenge on the Thai authorities and look for new economic resources by raiding Thai villages.

Although the UN deployed military observers to monitor the border, it was largely prevented from entering Khmer Rouge-controlled territories. Furthermore, the monitoring did not involve policing. Despite its 22,000 personnel, the UN stressed its lack of manpower to exert control and the difficulties it faced in obtaining information. UN military personnel also noted that the majority of logging for export was conducted by senior officials, including the governors of border provinces and high-level officers in each Cambodian faction, but without the official approval of their political authorities (Le Billon, 1999). To try to stop this would require large-scale military intervention by UNTAC, which would likely prompt armed resistance. Global Witness collected evidence through undercover investigations when an investigator posed as a timber buyer. When the UN obtained evidence of the ban being violated, such as video footage taken from a helicopter, the special representative of the Secretary-General decided not to make it public to avoid a confrontation with Thailand. Instead, the UN Secretary-General personally took up the matter with Thailand's Minister of Foreign Affairs, but to no avail.[9] Furthermore, Thailand was not the only country to violate the embargo, nor were all exports illegal, as the CPP provided export authorizations after the embargo was imposed (Le Billon, 1999).

Overall, the ban was relatively effective, but its short, 6-month duration and the lack of control over Khmer Rouge exports undermined its overall impact. In the end, Thai timber imports for 1993 were only 20% lower than they were in 1992 and 1994, and while Khmer Rouge revenues were certainly reduced, they were not completely curtailed. The unintended effects of the export ban included an increase in exploitation during the 3 months preceding its implementation, and the further criminalization of the timber trade.

With the formation of an uneasy, but internationally recognized coalition government bringing together the CPP and FUNCINPEC, the forestry sector rapidly evolved under the impetus of increasing demand and greater involvement by regional companies. This growth took place simultaneously with a complex post-UN transition process whereby FUNCINPEC saw the CPP more as an enemy than as an ally. On the one hand, the new government of Cambodia remained militarily opposed by the Khmer Rouge, but could not count on direct foreign assistance to help finance the ongoing war. But on the other, the plight of the population, and the guilt accumulated by donors guaranteed a sustained and relatively unconditional provision of humanitarian and development aid representing approximately half of the national budget (Grube, 1998). Despite this development aid, the leader of FUNCINPEC and new First Prime Minister, Prince Norodom Ranariddh, had few means and little resources to construct a power base

[9]The government of Prime Minister Chuan Leekpai (September 1992–1995 and 1997–2001) succeeded a military dictatorship installed by a coup in February 1991 and was initially not keen in directly confronting the economic interests of some military units.

within a state apparatus overwhelmingly controlled by the CPP. FUNCINPEC could either strengthen public support through better fiscal and institutional management practices, thereby improving the lot of civil servants and the population (thanks to policies developed by some of its key technocrats), or it could attempt to tap into some of the growing extra-state financial flows generated through corrupt practices already well entrenched by the CPP.

Both Ranariddh and Hun Sen opted for the latter option, agreeing on a mutually profitable sharing of resources. Once this choice was made, its logic became self-reinforcing: "since no progress had been made in creating a rule of law or building democratic institutions, the only way to protect one's wealth and position was to hold onto power" (Ashley, 1998, p. 69). In short, while both parties wanted to attract foreign investors, none had a real desire to strengthen the state to the detriment of its own clientele and factional interests. This choice was reflected in the incorporation of politically aligned civil servants into the government, as well as in the free reign given to a number of officials and military officers. In many provinces, provincial governors and generals acted as warlords, using violence to assert their control over forested areas and transportation routes, and flouting forest and tax regulations to enhance personal profit, or simply to finance both their administration and soldiers. While tax exemptions were widely granted to businesses, the two competing parties set up parallel "shadow" taxation schemes. Some of the slush funds that were generated resulted in the construction of "Hun Sen" or "Ranarridh" schools; however, others were allocated to the private interests and security of high-ranking officials (Le Billon, 1999). FUNCINPEC was unwilling to rely on a formal administration still controlled by the CPP, while the CPP was unwilling to credit FUNCINPEC for a governmental success. Furthermore, at every level of the government administration, civil servants had to struggle along on salaries well below subsistence, which only further promoted widespread corruption. The public treasury received on average of only 5% of the value of exported timber, well below the 12% internationally recognized as a minimal public rent (Le *Billon*, 1999, 2000). Ironically, the highest tax revenues were achieved when, with the knowledge of World Bank officials, former FUNCINPEC Minister of Finance, San Rainsy, succeeded in imposing taxation on logs exported from Khmer Rouge controlled territories (Le Billon, 1999).

As a result of this governance driven by factionalism and personal interests, uncontrolled practices in the logging sector flourished as the rapidly expanding private sector sought to benefit from the opening of this new frontier of capitalism. A large number of logging concessions and export licenses were granted without public announcement or bidding, and taxation on logging was frequently waived to increase shadow taxation and profits. Illegal logging increased sharply, alongside the award of concessions covering all available productive forests (Le Billon, 1999). As the government required the processing of timber before export, heavy investment was demanded from concessionaires. The absence of logging quotas, and the large surplus capacity in view of the sustainable harvest potential, resulted in illegal logging within the large protected areas set up by Royal Decree in 1993. The embezzlement of forest revenue largely benefited the army, which had retained de facto territorial control over that of the Forest Department, due to the ongoing war against the Khmer Rouge. In 1994, the co-prime ministers briefly put the army in a monopoly position of licensing timber exports (Royal Government of Cambodia, 1994). This logic of political accommodation to further factionalist economic interests was even extended to the Khmer Rouge, as the co-Prime Ministers approved log exports by Thai companies from territory controlled by the insurgent group (Global Witness, 1995b; Royal Government of Cambodia, 1996). By doing so, the government was providing a de facto seal of approval for the funding of the Khmer Rouge while the war went on.

This supportive relationship between supposed war enemies was repeated in numerous instances at the local scale. While outright confrontations did occasionally take place over control of rich forest stands, transportation routes, and log yards, commanders from both sides often came to mutually beneficial business agreements while maintaining a minimum level of conflict, thus guaranteeing their continued control of the region and its civilian institutions. In more simple and frequent cases, Khmer Rouge units were simply paid protection fees by companies or government army units involved in logging (Le Billon, 1999). This source of support proved crucial to front-line Khmer Rouge forces located far from their rear bases. While on the one hand this revenue fuelled the conflict, on the other, the accommodating political economy of war provided the opportunity for multiple contacts and common interests on both sides of the political divide, as well as a rapid *embourgeoisement* of local military commanders. These developments among certain Khmer Rouge cadre seem to have facilitated, if not induced, their defection in 1996, when the flow of timber was reduced because Thailand closed its border to Khmer Rouge exports in mid-1995 following international pressure by the British NGO Global Witness and the US government.

Despite the mutual accommodation of the elite, the political situation severely degraded from early 1996 onward. FUNCINPEC could not consolidate its political victory following the UN-sponsored elections, while the CPP increased its control of both the military and the state apparatus (Peou, 2000). The situation was complicated by the massive mid-1996 defections of Khmer Rouge following the evaporation of logging revenues.[10] The defections, however, were not general and local Khmer Rouge commanders remained in partial control of their territory and its resources through yet another accommodation with key governmental officials. The government coalition collapsed during a coup in July 1997 as a result of the instability created by the problem of allegiance over the newly available forces and the resources that they controlled. Since the coup, financial demands for the forthcoming electoral campaign and by the military, in part to conduct offensives against remaining royalist forces, led to more logging and

[10]The closure of the border was not the only factor pushing Khmer Rouge cadre to defect. Other factors included governmental military pressure, growing dissent from local commanders against the orthodox policies imposed by the party elite, and their unfulfilled demand for increased revenue transfer. The RGC had also secured the support of Thai companies operating in Khmer Rouge territory through a series of agreements with the Thai government asking companies to pay taxes to the Cambodian government in exchange for legitimacy (Global Witness, 1996).

timber exports, in particular towards Vietnam (Global Witness, 1998). New elections in July 1998 reconfirmed the CPP's grip on power, although the electoral victory failed to yield an absolute ruling majority. Thus a second coalition government was formed, this time with FUNCINPEC serving as the lesser partner.

With a legitimized hold on power, the CPP—and more specifically Hun Sen—did not have to rely systematically on a parallel budget to sustain its power base. Rather, the new government needed to repair its damaged image following the coup, and to consolidate the election dividend by responding to the demands of international donors. Topping these demands was the call for the clean up of the forestry sector, which the government addressed through a heavy-handed crackdown on so-called illegal activities. However, the target of this cleansing was mostly small loggers and unruly political clients, rather than key concessionaires, resulting in the loss of thousands of jobs. As such, the discourse of legalization and cleaning-up of the forestry sector legitimated a concentration of profits by condoning the monopolization of access to timber and the use of violence against small-scale illegal loggers, many of who were self-demobilized soldiers and impoverished peasants.

Despite significant progress in the independent monitoring of the forestry sector, conducted by the Forest Crime Monitoring Unit funded by the British government and staffed by Global Witness, the sector remained one characterized by crony capitalism and an economic rationalization and concentration of profits. At the end of 2001, following sustained pressure from donors in the light of renewed massive flooding, low tax revenues, and failure by forest concessionaires to provide management plans, Prime Minister Hun Sen suspended all concessions and prohibited logging, arguing that he preferred receiving the support of international donors rather than relying on logging revenues threatening the environment.

The official government stance was that illegal logging was considered to be exclusively small-scale timber theft (Global Witness, 2002; Le Billon, 2002). Law enforcement was not targeting organized forestry business, and in December 1999 Global Witness reported that Cambodia's major forest concessionaires were still engaging in illegal forest practices. Global Witness (2001) suggested that the concessionaires acted with effective impunity, citing the government's lack of consistency with regard to forestry reform as a key problem. Concomitant to such impunity are frequent allegations that those companies tied to top government ministers, senior military officers, and Hun Sen's relatives, are also those who are practicing illegal logging (Sipress, 2003).

In July 2002, Cambodia's National Assembly enacted a new forestry law making it illegal to cut trees outside concession areas, in national parks, in wildlife sanctuaries or in other designated areas. The legislation imposes heavy penalties of up to 10 years in jail and fines of up to USD 25,600 for illegal logging (Borton, 2002). It is difficult, however, to take such legislation seriously, given the well-documented weakness of Cambodia's judiciary, and the country's historical penchant for impunity and government corruption.

In late 2002, Global Witness, the most effective critique of government corruption and impunity related to Cambodia's forestry sector, became embroiled in a standoff with the RGC. In early 2003, Prime Minister Hun Sen had reportedly moved to sue Global Witness employee, Eva Galabru, for slander and has publicly stated that he intends to fire the group as the country's official forestry monitor (Sipress, 2003). The World Bank threatened, in response, to suspend its entire USD 20 millions aid program for Cambodia unless the government allows a continuing role for Global Witness. Because illegal forestry is so lucrative, many foreign diplomats in Phnom Penh were skeptical about Cambodian official interest to heed the World Bank's warning. Human Rights Watch (2002) also accused the government of harassing and intimidating Global Witness and environmental advocates.

Despite these warnings and critiques, the RGC sacked Global Witness in early 2003, although the NGO continued to operate in Cambodia as an unofficial forest monitor and publish damning reports (Coren, 2003; Global Witness, 2004). In November 2003, the World Bank negotiated the hiring of a new purportedly weaker forestry monitor, Swiss firm Société Générale de Surveillance (SGS), and the Bank set to disperse a new structural adjustment credit (SAC) worth US $15 million. This new deal has been severely criticized, as the terms of reference of the new deal have been much watered-down. For example, a clause requiring that the forestry monitor be free of "direction, control or influence of any ministry that will be audited or reviewed" was eliminated, a compulsory 6-month assessment of the government's compliance with its obligations was removed, and the random unannounced spot checks that regularly turned up evidence of illegal logging for Global Witness were also dropped under the new deal (Coren, 2003). In 2005, 10 years after the beginning of their campaign, Global Witness staff was finally made "persona non grata" in Cambodia by the government.

3.3 Logging in War and Transition to Peace

Forests have been significant for several reasons in the process of transition from war to peace that has taken place over the last decade in Cambodia. The first reason is geopolitical. Most forests are located at the periphery of the country, along the borders of Thailand and Vietnam, two countries that have often supported opposing Cambodian factions in order to increase their respective influence over Cambodia. Forests have thus a long history of playing a role in insurrections and politics, by providing a natural fortress or refuge for opponents to central authorities, and a safe conduit to the arms, equipment, drugs, or food provided by neighboring sponsors.

Second, the logging revenues controlled by politico-military factions became particularly important as the financial backing of foreign sponsors declined in the late 1980s and social status in society became increasingly tied to financial wealth. By helping to fund the conflict, the commodification of forests thus played an active role in its continuation. Again, the location of forests was important for providing a spatial continuity between the site of struggle and the source of support (market), in this case, by enabling Thai companies to gain access to Cambodian forests controlled by particular factions. The capture of this economic rent also expanded military objectives to include the control of forested areas and transport routes. Opportunistic local commanders and politicians imposed their rule, often violently, upon local populations and forestry officials. In this way, the behavior of the government army in the 1990s was not very different from that of the early 1970s, when the generals and troops of former military ruler Lon Nol used the wide-ranging power provided to them by the regime and the context of war to log and export timber.

A third reason why forests have been significant in Cambodia's transition to peace is the socio-political effect that this commodification had on actors organizing and controlling the commodity network. More than money, logging provided networks of support reaching important military and political figures in Thailand and Vietnam. Most prominently, former Thai Prime Minister Chavalit and other Thai politicians used bribe revenues to enhance their electoral slush funds. In Vietnam, the collusion of the provincial ruling elite and the absence of transparent government or civil institutions also provided an enabling regional environment. Direct contacts developed at the local level between military commanders, politicians and businessmen also nurtured the personal economic interests of members of the factions as well as co-operative arrangements between ostensible enemies. These socio-economic relations diluted the orthodoxy of Khmer Rouge individual political commitments and weakened the general chain of command of military factions. This was particularly true when the commodity network crossed a political divide, either spatially (e.g., an export route successively controlled by opposing factions) or institutionally (e.g., the need for an export license required by the central authorities), and thus demanded accommodation by the different sides.

A fourth reason is the availability of windfall revenue from large-scale logging operations had a significant politico-economic effect on the structure of power in a country as impoverished as Cambodia. By providing a means for patrimonialism and opportunities for corruption, the forestry sector weakened formal governance, reducing it to profit taking by individual and factional interests. When the transition to peace occurred, the local effect of forest commodification also played a significant role in structural development of the polity, as it provided opportunities for peripheral actors, mostly provincial strongmen and politicians, to benefit from the conjuncture of economic liberalization and political uncertainty to strengthen their power base by increasing their financial wealth and building local production networks. In turn, the central leadership counteracted this decentralization of resource control and political power with a concessionary framework ensuring a more secure control and capture of forestry rents through foreign companies or cronies.

A fifth link of the role of forests in the peace process is the antagonism that the mode of forest exploitation created between the government of Cambodian and the local population, as well as the international community. The local population and international community perceived the government of Cambodia as corrupt, and the administration of the forestry sector reinforced this impression. Much of the antagonism between these various actors resulted from the confrontation of pragmatist and idealist perspectives regarding the transition process itself. In this regard, the rationalization of forestry through logging concessions accompanied by improvements of governance, economic indicators, and the accountable collection of public revenue by the government diminished this antagonism and won back a degree of support from the international community for a few more years. The international community played a dual role through its discourse on sustainable and accountable resource management, as it enabled open criticisms of skewed and illegal practices, but also pushed practices further underground. The tenets of a "green" reconstruction were deeply contradictory given the realpolitik at play in Cambodia. The apparent failure of the government to control and effectively tax logging can be interpreted as resistance to the demands of the international community for greater apolitical, or nonpartisan, development of resources. Not only a tactic of resistance however, illegal logging was used by politicians and military to line their pockets and further their ultimate objective of regime and individual security. Faced by such conditions for reestablishing governance—strict but poorly enforced logging and taxation guidelines—neither of the co-prime ministers had a strong incentive in consolidating the rule of law and public services, beyond deceptive pronouncements, if it was to the detriment of their own clientele and particular interests. The international community's public transcript of legalization and rationalization of the forestry sector also entrenched the legitimacy of foreign transnational companies until adequate independent monitoring demonstrated that they were no better than local loggers.

Finally, the Cambodian example suggests that neoliberal prescriptions can be accommodated and redirected by state elites in ways that have profoundly negative implications for the general population. While politico-military elites and their circle of clients have benefited and continue to profit from Cambodia's logging sector, the population has seen its access to timber resources and revenues drastically reduced. Cambodia remains one of the least developed countries in the world, even though hundreds of millions of dollars have been generated from logging in the last decade, and the plunder of the forests has continually been justified in the name of development (Bottomley, 2002). The failure of democratization in the 1990s played a role in this plunder as ruling elites remained largely unchallenged. The main responsibility of international agencies dealing with logging issue was to emphasize an industrial/corporate approach rather than a community-oriented one. In combination with biased neoliberal reforms and the imperative of drawing fiscal revenues from forests, this approach came to accommodate and ultimately serve the political and economic interests of a tiny minority, with disastrous consequences for Cambodia's environment and the livelihood of the population at large.

4 CONCLUSION

Examining the incorporation of forests into the transition of the 1990s, two distinct aspects dominated: forests as source of revenue and reward for political supporters; and forests as space of social escape and political resistance. From the early 1990s, these two aspects were combined as opposing Cambodian factions commodified forests for the sake of political struggle. Forests, perceived as part of the benefits of war, protracted the conflict as privileged access to timber was gained through violence. However, such control also rested on compromises and co-optation motivated by the multiplicity of interests associated with logging and the practical challenge of bringing timber to market.

As such, the rational of this political economy of war lowered the intensity of violence and provided the opportunity of multiple contacts and the development of common interests on both sides of the political divide as well as a rapid *embourgeoisement* of local military commanders. These developments among certain Khmer Rouge units seem to have facilitated, if not induced their defection in 1996. In this way, the analysis of an economic logic of war requires a careful appraisal of the relations associating actors through both material and discursive resources at a variety of scales.

As such, the categories of peace, war, ally and enemy were somewhat flawed in the context of Cambodia and the concepts of accommodation, facade politics, and fluid productive networks were more fruitful in bringing forward the logic and socio-political intelligibility of logging, and more generally those of any benefits of war. Broadly speaking, logging and the evolution of governance were mutually related. As governance rested upon unaccountable shadow practices, and the flow of timber was ultimately controlled at the local level, commodity networks occasionally escaped central control and took on a life of their own. In this way, logging turned supporters into unruly clients and even strongmen, competing with the central leadership of the various factions. This was particularly the case when a timber export ban was imposed against the will and interests of local actors on both sides of the border. On the other hand, by authorizing the flow of timber to take place, the central leadership could accommodate these interests and place itself in the position of patron. Although forests became a space of resistance to the practices of the Cambodian regime, and the center of attention and contention between international development actors and bureaucratic elites, the central leadership successfully enrolled the international development agenda prescribing sustainable logging, a prescription that proved all but self-illusory given the political, economic, and institutional contexts of the country. The pressure of international donors and civil society groups did bring about a greater degree of transparency and accountability from governments, businesses, and importing countries through a mix of aid conditionality, technical assistance, and economic sanctions. The accommodation of the central elite to these requirements, however, proved only temporary. Rather than pursuing from the onset a radical attempt at consolidating forest access for local communities, the international community ultimately ended up consolidating the "shadow" practices of a predatory state.

REFERENCES

Asian Development Bank (ADB). (1991). *Economic Report on Cambodia*. Manila: Asian Development Bank.

Ashley, D. W. (1998). The failure of conflict resolution in Cambodia: Causes and lessons. In F. Z. Brown & D. G. Timberman (Eds.), *Cambodia and the International Community* (pp. 49–78). Singapore: ISEAS.

Bates, R. H. (1981). *Markets and States in Tropical Africa: The Political Basis of Agricultural Policies*. Berkeley, CA: University of California Press.

Bilgin, P., & Morton, A. D. (2002). Historicising representations of 'Failed States': Beyond the cold-war annexation of the social sciences? *Third World Quarterly, 23*(1), 55–80.

Borton, J. (2002, Aug 30). Reform in the forests. *Asia Times Page*, from http://www.atimes.com/atimes/Southeast_Asia/DH30Ae01.html.

Bottomley, R. (2002). Contested forests: An analysis of the highlander response to logging, Ratanakiri Province, Northeast Cambodia. *Critical Asian Studies, 34*(4), 587–606.

Broad, R. (1995). The political economy of natural resources: Case studies of the Indonesian and Philippine forest sectors. *Journal of Developing Areas, 29*(3), 317–340.

Broad, R., & Cavanagh, G. (1993). *Plundering Paradise: People, Power and the Struggle over the Environment in the Philippines*. Berkeley, CA: University of California Press.

Brown, F., & Timberman, D. G. (Eds.) (1998). *Cambodia and the Internal Community: The Quest for Peace, Development and Democracy*. New York: Asia Society and ISEAS.

Bryant, R. L. (1997). *The Political Ecology of Forestry in Burma 1824–1994*. London: Hurst.

Chandler, D. (1996). *A History of Cambodia*. Boulder, CO: Westview Press.

Colchester, M. (1993). Pirates, squatters and poachers: The political ecology of dispossession of the native peoples of Sarawak. *Global Ecology and Biogeography Letters,* 3, 158–179.

Coren, M. (2003, Nov 21–Dec 4). New witness of illegal logging poised to sign with government. *Phnom Penh Post, 12*(24), from http://www.phnompenhpost.com/.

Dauvergne, P. (1997). *Shadows in the Forest: Japan and the Politics of Timber in Southeast Asia*. Cambridge, MA: Massachusetts Institute of Technology.

Food and Agriculture Organization (FAO). (1994). *Cambodia Land Cover Atlas 1985/87–1992/93*. Phnom Penh: Food and Agriculture Organization.

Food and Agriculture Organization (FAO). (1997). *State of the World's Forests*. Rome: Food and Agriculture Organization.

Findlay, T. (1995). *Cambodia: Legacy and Lessons of UNTAC*. Oxford: Oxford University Press.

Ghee, L. T., & Valencia, M. J. (Eds.) (1990). *Conflict over Natural Resources in South-East Asia and the Pacific*. Tokyo: United Nations University Press.

Global Witness. (1995a). *Forests, Famine and War: The Key to Cambodia's Future*. London: Global Witness.

Global Witness. (1995b). *Thai-Khmer Rouge Links and the Illegal Trade in Cambodia's Timber*. London: Global Witness.

Global Witness. (1996). *Corruption, War and Forest Policy: The Unsustainable Exploitation of Cambodia's Forests*. London: Global Witness.

Global Witness. (1998). *Going Places ... Cambodia's Future on the Move*. London: Global Witness.

Global Witness. (2001). Global Witness. *The Credibility Gap and the Need to Bridge It: Increasing the Pace of Forestry Reform*. London: Global Witness.

Global Witness. (2002). *Deforestation Without Limits: How the Cambodian Government Failed to Tackle the Untouchables*. London: Global Witness.

Global Witness. (2004). *Taking a Cut*. London: Global Witness.

Graf, W. (1995). The state in the third world. *The Socialist Register, 31*, 140–162.

Grube, D. I. (1998). *Donors in Disarray: Prospects for External Assistance to Cambodia*. Phnom Penh: Cambodia Development Resource Institute.

Heder, S., & Ledgerwood, J. (Eds.) (1996). *Propaganda, Politics, and Violence in Cambodia: A Democratic Transition under United Nations Peace-keeping*. New York: M.E. Sharpe.

Hirsch, P. (1995). Thailand and the new geopolitics of southeast Asia: Resource and environmental issues. In J. Rigg (Ed.), *Counting the Costs: Economic Growth and Environmental Change in Thailand* (pp. 235–259). Singapore: Institute of Southeast Asian Studies (ISEAS).

Hughes, C. (2003). *The Political Economy of Cambodia's Transition, 1991–2001*. London: Routledge.

Human Rights Watch. (2002, Dec 24). *Cambodia: Excessive Force Used Against Environmental Advocates*, from http://www.hrw.org/press/2002/12/cambodia1223.htm.

Hurst, P. (1990). *Rainforest Politics: Ecological Destruction in South-East Asia*. London: Zed Books.

Le Billon, P. (1999). *Power Is Consuming the Forest: The Political Ecology of Conflict and Reconstruction in Cambodia*. D. Phil Thesis. University of Oxford, Oxford, UK.

Le Billon, P. (2000). The political ecology of transition in Cambodia 1989–1999: War, peace and forest exploitation. *Development and Change*, *31*(4), 785–805.

Le Billon, P. (2001). The political ecology of war: Natural resources and armed conflicts. *Political Geography*, *20*(5), 561–584.

Le Billon, P. (2002). Logging in muddy waters: The politics of forest exploitation in Cambodia, *Critical Asian Studies*, *34*(4), 563–586.

Ministry of Economy and Finance. (1997). *Economic Outlook for Cambodia 1997–2001*. Phnom Penh: Ministry of Economy and Finance.

Ministry of Environment. (1994). *Cambodia: First State of the Environment Report 1994*. Phnom Penh: Ministry of Environment.

Morello, T., & Thayer, N. (1992, Nov 26). The Pol Pot trail: UN efforts to end Khmer Rouge trade likely to fail. *Far Eastern Economic Review*.

Neiss, H. (1997). *Statement by Hubert Neiss, IMF Representative*. Paris: Consultative Group Meeting.

Peou, S. (2000). *Intervention and Change in Cambodia: Towards Democracy?* New York: St. Martin's Press, Institute of Southeast Asian Studies & Silkworm.

Potter, L. (1993). The onslaught of the forests in South-East Asia. In H. Brookfield & Y. Byron (Eds.), *South-East Asia's Environmental Future: The Search for Sustainability*. Kuala Lumpur: UNU Press and Oxford University Press.

Regaud, N. (1992). *Le Cambodge dans la Tourmente: le Troisième Conflit Indochinois, 1978–1991*. Paris: L'Harmattan.

Reno, W. S. K. (1995). *Corruption and State Politics in Sierra Leone*. Cambridge, UK: Cambridge University Press.

Royal Forest Department. (1996). *Forestry Statistics of Thailand*. Bangkok: Royal Forest Department, Government of Thailand.

Royal Government of Cambodia. (1994). *Decision on the Annulment of the Former Procedures and Regulations on the Export of Timber*. Phnom Penh: Royal Government of Cambodia.

Royal Government of Cambodia. (1996). *Letters to Banharn Silpa Archa Authorising Log Exports to Thailand*. Phnom Penh: Royal Government of Cambodia.

Scott, J. C. (1990). *Domination and the Arts of Resistance: Hidden Transcripts*. New Haven, CT: Yale University Press.

Sipress, A. (2003, Feb 16). Illegal Logging Rife in Cambodia, *Washington Post*, p. A26.

Supreme National Council. (1992). *Declaration on the Export of Logs from Cambodia, 22 September 1992*. Phnom Penh: Supreme National Council.

Talbott, K. (1998). Logging in Cambodia: Politics and plunder. In F. Z. Brown & D. G. Timberman (Eds.), *Cambodia and the International Community* (pp. 149–168). New York: Asia Society.

Thayer, N. (1991, Feb 7). Rubies are rouge: Khmer war effort financed by gem finds, *Far Eastern Economic Review*.

United Nations Development Programme (UNDP). (1989). *Report of the Kampuchea Needs Assessment Study*. New York: UNDP.

Vitug, M. D. (1993). *The Politics of Logging: Power from the Forest*. Manila: Philippine Center for Investigative Journalism.

World Bank. (1992). *Cambodia: Agenda for Rehabilitation and Reconstruction*. Washington, DC: East Asia and Pacific Region Department, The World Bank.

World Bank, UNDP, and FAO. (1996). *Cambodia, Forest Policy Assessment*. Washington, DC: World Bank Group.

CHAPTER 3

GREED OR GRIEVANCE IN WEST AFRICA'S FOREST WARS?

RUBEN DE KONING

1 INTRODUCTION

After the end of the Cold War and the improvement of relations between the super powers, violent conflicts continued to be part of Africa's political reality. Between 1992 and 2001, the number of armed conflicts outside Africa dropped by half, yet in Africa the number stayed roughly the same (Ross, 2003). Richards (2001, p. 65, referring to Duffield, 1998) identified a new type of "forest war" on the rise in the western half of the continent, stretching from Zaire to Liberia. These wars, largely fought over and sustained by natural resources, signify a break with old conflicts which were largely rooted in positions created by Cold War animosities. Apart from some exceptions such as Angola, such proxy conflicts (some of which continue until today) were located on the eastern side of the continent, from the Horn of Africa to Mozambique.

Insurgence movements in countries like Liberia, Sierra Leone and Zaire emerged in areas that contained spatially concentrated, high-value resources such as diamonds, gold and other precious metals (Deacon & Mueller, 2004, p. 28). There is considerable evidence that the sales of such resources have enriched insurgents and provided them with the financial means to wage war (Global Witness, 2003, 2004b; UNSC, 2000, 2001, 2004). Referring to ongoing conflict in central Africa, Reno (2000a, p. 4) argues that warfare is to be understood as "an instrument of enterprise and violence as a mode of accumulation." The strong correlation between abundant natural resources and civil war observed in Africa but also elsewhere, leads analysts to conclude that economic greed should be recognized as the motivation of civil war today, rather than politically or ethnically inspired ideologies or grievances.[1] Collier (2000, p. 92) argues that "since both greed motivated and grievance motivated rebellions will embed their behavior in a narrative of grievance, the observation of that narrative provides no informational content to the researcher as to the true motivation for rebellion."

[1] By analyzing the relative significance of variables that proxy grievance and greed, using a new data set of wars during 1960–99, Collier and Hoeffler (2001) argue that economic agendas outperform any other.

W. de Jong, et al. (eds.), Extreme Conflict and Tropical Forests, 37–56.

The proposition that resource abundance causes civil wars by providing incentives for greed based insurgence runs directly counter to the environmental security argument that gained prominence in the field of conflict studies in the mid-nineties (see Baechler, 1999; Homer-Dixon, 1994, 1999). Homer-Dixon, an exponent of this perspective, foresees intrastate conflict resulting from resource scarcities as opposed to resource abundance. Environmental scarcities are the result of degradation, increased demand and unequal resource distribution. When such processes are severe—often as a result of rapid population growth—they lead to unequal resource appropriation, forced migrations, and ecological marginalisation of powerless groups. Out of desperation these vulnerable groups are forced to fight, either with each other over local resources or with powerful elites.

There is little evidence that scarcity theory applies to forest wars of Central and West Africa. Although many African countries are now often disturbed, degraded, and destitute, those countries in which war has erupted were no more degraded before conflicts started than their more peaceful African counterparts. The "economic agendas" framework seems more appropriate to understanding violent conflict in this part of the world; indeed many recent country case analyses depart from such understanding (Atkinson, 1997; Keen, 2003; Le Billon, 2001; Reno, 1997, 2000b; Vlassenroot & Romkema, 2002). Although it is evident that resource exploitation has played a major role in perpetuating violent conflict in Central and West Africa, we should be cautious in claiming that resource abundance *causes* violent conflict. While wars may involve resource mobilization, environmental factors in and of themselves do not cause social phenomena such as violent conflict and are only one aspect of the context in which these phenomena evolve. Human groups must interpret environmental preconditions, like any other precondition, and either accept or reject them before they respond in violent or nonviolent ways. Libiszewski (1992, p. 13) rightfully comments that "in social science research and particularly in conflict research we should understand causes in the sense of reasons". Yet this is exactly what is ignored by Collier and associates when they use positive correlations between resource abundance/dependence and civil war to validate claims about people's motives for engaging in violent conflict.

This chapter argues that because violent struggles in West Africa evolve around natural resources does not mean that insurgents' motives are necessarily greed based. When regarding the historical particularities and rebel narratives of civil wars in Liberia, Sierra Leone and Ivory Coast, we see that strong political and social grievances surface that provided important motives for violent struggle. In fact socio-political grievances easily meld with economic opportunism in the praxis of warfare. Based on the case studies, this chapter criticizes the way Collier's econometric study of civil war sets and understands proxy variables for greed and grievance based motivations. The chapter argues that isolating and stressing greed as the principal motivation for rebels is confusing and may impede efforts to end conflict. Firstly, the greed narrative steers governance responses towards clean interventions such as boycotts, travel bans, and export diversification. These do not involve strong military intervention in messy conflict situations, something that in many cases has proven to be badly needed. Secondly,

Table 3.1. *Acronyms used in the chapter.*

ECOMOG	The Economic Community of West African States Cease-fire Monitoring Group
FPI	Front Populaire Ivoirien
ICG	International Crisis Group
LURD	Liberians United for Reconciliation and Democracy (Liberia)
MJP	Mouvement pour la Justice et la Paix (Ivory Coast)
MODEL	Movement for Democracy in Liberia
MPCI	Mouvement Patriotique de la Côte d'Ivoire
MPIGO	Mouvement Populaire du Grand Ouest (Ivory Coast)
NPFL	National Patriotic Front of Liberia
NPRC	National Provisional Ruling Council (Sierra Leone)
RDR	Rassemblement des Républicains (Ivory Coast)
RUF	Revolutionary United Front (Sierra Leone)
ULIMO	United Liberian Movement for Democracy (Liberia and Sierra Leone)
UN	United Nations
UNAMSIL	United Nations Mission in Sierra Leone
UNMIL	United Nations Mission in Liberia

the greed narrative tends to criminalize insurgents and denounces any political and social agenda. A better understanding of people's perceived reasons for engaging in violent conflict can instruct more genuine and engaged interventions.

2 LIBERIA: A ONE MAN SHOW RIDING ON RURAL FRUSTRATIONS

On Christmas eve 1989 the National Patriotic Front of Liberia (NPFL)[2] (see Table 3.1) began incursions into Nimba County in northern Liberia from Ivory Coast (Ellis, 1999, p. 110). These were spearheaded by two political dissidents: Charles Taylor, who became the political leader of the rebel movement, and military strategist Prince Johnson. Both had been expelled from the country when President Samual Doe took over power in a violent coup in 1980. The core NPFL rebel movement consisted of dispersed Liberians, most of whom had found refuge in Ghana and Burkina Faso, and some of whom had received military training and support from Libya's leader Ghadaffi (Richards, 1996, p. 2).[3]

Although the NPFL forces initially counted only about 100 men, their number increased rapidly through the recruitment of local Nimba youths, who proved ready combatants in Taylor's war (Utas, 2003, p. 15). Generally the NPFL was widely welcomed by Nimba County civilians who had come to despise the Doe government's

[2] The National Patriotic Front of Liberia was formed in Burkina Faso by soldiers and supporters of the late Corporal Quiwonkpa, who was brutally killed by the Samual Doe regime for being alleged to have plotted a coup in 1985 (Ellis, 1995).

[3] Ghadaffi was profoundly anti-Doe, not because its reckless rule, but for military strategic reasons. Liberia under Doe is said to have been a base from which Americans undertook operations against Libya. Furthermore, Liberia kept close political and economic relations with Israel throughout the 1970s and 1980s (Richards, 1996, p. 2).

ruthless and violent measures to put down the rebellion (Richards, 1996, p. 3). Marching up to Monrovia the NPFL split into two sections, one following Johnson and the other following Taylor. Prince Johnson's group captured and killed Liberia's president, plunging the country into 6 years of anarchy, during which rebel factions fought each other and ECOMOG, the Economic Community of West African States Cease-fire Monitoring Group, first deployed in the country in 1990.

A peace agreement, signed in 1996, allowed presidential and legislative elections to be held in 1997, in which Charles Taylor won with a landslide victory. Stability, however, was still far away. Opposing factions including the Liberians United for Reconciliation and Democracy (LURD) and the Movement for Democracy in Liberia (MODEL), continued to organize themselves in peripheral forested areas of neighboring Guinea and Ivory Coast. When fights broke out in the streets of Monrovia in 2003, the national and international outcry forced Taylor to step down as president. A peace agreement was signed in August 2003, involving different political and military groups and civil society organisations who made up the transitional government led by Chairman Gyude Bryant. Disarmament and demobilization started in 2004 under the new United Nations Mission in Liberia (UNMIL).

The brutal and seemingly irrational violence and pillage that characterized civil war in Liberia was unprecedented and left the international community puzzled on how to understand, let alone respond to this new type of warfare. The most tangible stakes in this war appeared to be valuable diamond, gold and timber resources. Taylor seized and revived Liberia's timber industry before and during his presidency, leading to a large increase in log volumes and export earnings. From almost nothing, Liberia's timber industry increased to account for 50% of its national export earnings and 20% of its GDP in 1999 (Archibald *et al.*, 2005, pp. 6–7). The opposing factions, LURD and MODEL fighters, sustained themselves by extorting money for timber shipments at border crossings, low technology mining of gold and diamonds, and wildlife hunting for bushmeat (Global Witness 2004a). Some of these practices continue until today. Key figures in the transitional government were known warlords drawn from former rebel groups who acted as gatekeepers to natural resource sectors in the country (ICG, 2004b).

Civil war and contemporary politics in Liberia was and is to a large extent a contest of power between competing alliances to gain control of the nation or at least the resource rich regions. However, besides the proximate cause of economic greed that appears to underlie the Liberian conflict, a complex historical legacy of legal dualism, discrimination, and ethnic differentiation need to be taken into account to explain why and how young men and women rallied to one or the other movement.

The first factor shaping grievances in Liberian society is the legal duality of land ownership created on behalf of the former American slaves who settled in Liberia in the nineteenth century. Their land ownership was recognized through individual title deeds, while indigenous populations remained operating under local customary authority that recognized user rights only. This system persists today, although some educated rural people have been able to acquire land under title deeds recognized by the government (Archibald *et al.*, 2005). Liberia's American-Africans and other urban educated people are referred to as *kwii* (the civilized, those ruled by settler law), in opposition to a variety of interior indigenous populations, e.g., the Mano, Loma, Gio, Vai and Kru. Collective tenure systems based on rotating user rights may be valued for their sustainable and egalitarian notions, however, such arrangements are not always in harmony with people's aspirations in life, particularly for young people. As in many African societies, a large share of Liberia's indigenous youth found itself trapped between customary systems of redistribution—elders controlling transfer of land and bride wealth—and globally inspired imaginations of individual wealth and independence. The tragedy of the Liberian case is that inequalities in access to land and other properties still are confirmed by law.[4]

A second source of grievance stemmed from the privileged status of the Mandingo people under previous governments since 1944. The Mandingo migrated several hundred years ago from present-day Guinea to settle in northern Liberia. Ever since, Mandingo merchants have acted as middlemen between coastal habitants and powerful empires of present-day Mali, Senegal, and Guinea. Interior chiefs used war to thwart this trade, which impinged on their territorial control (Archibald *et al.*, 2005, p. 58). The historical animosity between the Liberian Mandingo and interior agricultural based ethnic groups became an important rift in Liberia's war. By attacking the privileged Mandingo communities, Taylor won the critical support of Mano and Gola populations and the allegiance of many of their young men. In later stages of the war, this strategy backfired when the Mandingo dominated United Liberian Movement for Democracy (ULIMO) and LURD rebel factions successfully contested Taylor's rule until his downfall in 2003. The factions consisted of Mandingo refugees that had been driven from their lands in Liberia to sanctuary in Sierra Leone and Guinea respectively. Once the security situation improved, Mandingo refugees returned to their villages claiming back their properties and their traditional niche in the regional trading system. Recent media reports indicate that tensions are rising again, particularly in Lofa and Nimba County, where Mandingo returnees compete with local Loma, Gio, and Mano communities and refugees for land (Gongloe, 2004).

When reading narratives of army commanders and ex-combatants in the Liberian civil war, elements of grievance frequently appear and become difficult to disentangle from greed-based motivations. One statement by a LURD combatant in Zorzor Town, Lofa County quoted in Archibald *et al.* (2005, p. 8) is illustrative:

> "The elections will change nothing; they will bring their brothers and sisters from America and plug them into the political system, then leave us aside to fight amongst ourselves. They think we are useless people, just floating people. We are not fighting for liberation anymore, only for jealousy, because the country is not for everyone, only some people enjoy it."

By using the word jealousy, it becomes difficult to pinpoint his motivation into either greed or grievance. There is a sense of injustice (grievance) that, when recognized and

---

[4] Only in regard to marriage and women's property rights has dualism been brought to an end. This has been the result of efforts of women lawyers and legislators (Archibald *et al.*, 2005).

acknowledged through a process of political manipulation and mobilization, give the impunity to use violent means to redress this situation, often in the form of looting. In this sense violent looting signifies more than the act of taking something you want; it is also an expression of perceived injustices.

3 SIERRA LEONE: YOUTH REBELLION FOUGHT WITH DIAMONDS

The beginning of civil war in Sierra-Leone is usually traced back to March 23, 1991 when a small contingent of guerrillas, calling itself the Revolutionary United Front (RUF), entered the country from Liberia through the Gola forest belt (Abdullah and Muana, 1998 p. 147). The RUF consisted of Sierra Leonean exiles together with Burkinabé and Liberian mercenaries lead by former army corporal Foday Sankoh. Like the NPFL from Liberia, the movement was supported by Burkina Faso and Libya through training and armaments. The movement was driven by Charles Taylor, who wished to gain access to diamond-rich areas in Sierra Leone and topple the country's government that supported the ECOMOG peace-keeping mission that curtailed Taylor's aspirations in Liberia.

Before the RUF could reach Freetown, an urban based group of military officers staged a coup in 1992, installing the National Provisional Ruling Council (NPRC). With the help of anti-Taylor ULIMO rebels the NPRC forced back the RUF. Severely weakened, the RUF changed tactics and retreated to bush camps in the forested east of the country, from which it carried out extremely violent hit-and-run attacks throughout the 1990s in all parts of the country, including Freetown (Richards, 1996). The RUF kept itself alive by seizing important diamond mines, much to the annoyance of the NPRC government that was struggling to repay its international debts (Fairhead, 2001). In 1995, the NPRC hired the South African security firm, Executive Outcomes, for military assistance, paid with profits from diamond mines under government control (Reno, 1997).

The struggle for control of diamond-rich areas continued for years until British Special Forces directly intervened in 2000. The ineffective ECOMOG forces were replaced by the United Nations Mission in Sierra Leone (UNAMSIL). Sankoh was detained and elections took place in 2002. The RUF was transformed into a political party and elections reinstated Ahmed Tejan Kabbah to the presidency to which he had been elected in 1996. This change has brought relative stability until just recently.

Access to diamonds and diamond trading opportunities were reoccurring features of the civil war in Sierra Leone. Some commentators go as far as to suggest that the greed for diamonds overrides all other causes of conflict. Sierra Leone's ambassador to the UN, Ibrahim Kamara, (quoted in Crossette, 2000, p. 9), commented:

> "We have always maintained that the conflict in Sierra Leone is not about ideology, tribal or regional differences. It has nothing to do with the so-called problem of marginalized youths or . . . an uprising by rural poor against the urban elite. The root of the conflict is diamonds, diamonds and diamonds."

Kamara is right about his first claim, that is, that ideology, tribal or regional differences did not fuel civil war. The two major ethnic groups (Mende in the south and Temne in the north) each had their own political party and distinct sphere of influence. A third significant group consists of descendants of the liberated African population of the capital, known as "creoles". Although creoles dominated the public services they could not gain the same political preeminence as the Americo-Liberian community in Liberia (Clapham, 2003, p. 9). The absence of significant interethnic tension at country level was reflected in RUF bush camps. Communitarian markers were rejected, native languages banned and Christianity and Islam were considered equal (Richards, 2000, p. 166).

Kamara's second and third claim, that there is no youth or a rural-urban side to civil war, should be reexamined however. Rich evidence gathered by Richards (1996, 2000, 2001 and 2003) support the claim that the RUF in its early days certainly had a social and political programme that was inspired by the marginalized position of rural youth. In a political pamphlet, *Footpaths to Democracy*, written in 1995, the RUF for the first time spelled out its desires and beliefs, giving an idealistic representation of village life according to egalitarian RUF ideals and based on the provision of elementary services. Video footage of actual RUF bush-camps confirms this ideal showing neat lines of sleeping huts, complete with mosque and church inside a defensive ring of forest (Richards, 2001, p. 79). According to Richards, the RUF thus respond to the desire of outcast diamond diggers and other homeless youths to build secure homes in a community in which people are equal. Ideal camp life represents an alternative to, or rather an inversion of, the so much despised conditions in rural shanty towns. In the shanty towns, local strongmen and traditional chiefs exploit their people, living conditions are uncertain and violent, and people have no access to communications, schools, hospitals and roads. The allegiance of young diamond diggers and other rural youths to the RUF must be read as a complaint against the corrupt political and economic system blamed for the suffering of people in the nation's forested periphery.

Other analyses of the youth factor have concentrated on the emergence of a movement of urban semieducated youngsters in Freetown and abroad (McIntyre and Addo 2002; Richards 1996). The small group of exiles that comprised the initial core RUF rebellion had the common experience of being politically excluded under the All People's Congress of Siaka Stevens (1968–85) because of their political protest and student activism. Their ideas were much influenced by Ghadaffi's ideology of radical change as spelled out in his *Green Book* and the revolutionary spirit that circulated at the universities of Accra and to a lesser extent Abidjan.[5] *Green Book* communitarian ideals of community members' rights to basic needs and the devolution of power to people's assemblies were in line with RUF camp practices. Drugs, clothes and shoes—captured from government sources—were redistributed amongst rebels as well as captives, and

[5] The violent power takeover and successful transition to democracy in Ghana under the charismatic Jerry Rawlings inspired many young exiled students.

Sankoh's lack of presidential ambition stressed the movement's emphasis on collective decision making (Richards, 1996, 2000). In order to build a broad political constituency the core RUF group of intellectual exiles found not only a common cause with poor rural dwellers but also with school drop-outs and unemployed youths of Freetown, although these were equally inclined to join NPRC cadres. With a symbolic repertoire ranging from the *Green Book* to Rambo films, the RUF brought together a combination of urban and rural, uneducated and intellectual youth in one movement that aimed to change national and local social and political structures to the benefit of individual followers and captives.

Still loyal to its political and social project, RUF violence in its early days was mainly directed at those with a direct stake in the vested political and economic system, such as civil servants, government soldiers and traders. From the mid-1990s, however, the RUF more and more began to apply a scorched earth policy—looting and destroying villages, killing and mutilating its inhabitants—leaving commentators to conclude that a senseless primordial barbarity had taken over completely. This shift of tactics, however, can be explained (Richards, 2000). In 1996 a ceasefire and peace agreement was mediated by Ivory Coast, seemingly giving political recognition to the RUF movement. However, the reality on the ground proved different. The government hired international private security firms to defeat the RUF and regain diamond mines, while at the same time Nigerian peace-keeping forces detained RUF leader Fodah Sankoh and drove the RUF back into the forest. Realizing that political recognition at national levels was not attainable in the near future the RUF turned its hope again to the countryside, only to find that the balance had shifted against the movement through ethnically organized and government sponsored civilian vigilante groups (Richards, 2000). As it was considered legitimate to wage a war against a corrupted and exclusivist political regime, so was it now legitimate to target the civilians that stood in the way of the RUF cause and were seen as an extension of the political regime.

4 IVORY COAST: POLITICAL MANIPULATION OF CITIZENSHIP

Rapid economic decline in the 1980s after two decades of impressive economic growth, in combination with the death of independence president Houphouet-Boigny in 1985 set in motion a political struggle between prominent members of the late president's entourage over the remaining economic scraps. In this struggle president Henry Bédié (1993–99), was the first to employ ethnic narratives to play out his opponents. His notions of *Ivoirité*, ascribed to predominantly Christian southerners, went down well with large sections of society that felt that immigrants, grouped together with people born in the Maninka and Senoufou regions in the north, had profited disproportionately from economic growth and today had the remaining economic assets in hand. In December 1999, general Gueï staged a coup that removed Bédié from office, but adopted his xenophobic policies against Muslim northerners. In the run-up to elections in 2000 Gueï disqualified candidate Alassane Ouattara, who represented the northern-backed *Rassemblement des Républicains* (RDR), on the grounds that as a son of Burkinabé parents he was not a pure Ivorian. This led to the first violent skirmishes in Abidjan.

Despite this successful exclusion of "foreigners", it was political outsider Laurent Gbagbo who became president after his *Front Populaire Ivoirien* (FPI) supporters held massive protest against Gueï's self-proclaimed leadership. Expecting a coup from the side of Ouattara, Gbagbo's security forces randomly targeted his suspected supporters. In September 2002, just when relations seemed to improve between the two major political factions, a group of around 700 soldiers, united under the *Mouvement Patriotique de la Côte d'Ivoire* (MPCI), attempted a coup d'état (ICG, 2003). Unable to take Abidjan, the MPCI seized cities and towns in the northern and central regions, dividing the country in two parts. This division has remained intact until today. A neutral buffer zone between north and south was installed and is controlled by predominantly French troops.

The rise of insurgency and counterinsurgency in the western Ivory Coast has given more reason for concern in recent years. In November 2002 two opposition movements the *Mouvement Populaire du Grand Ouest* (MPIGO) and the *Mouvement pour la Justice et la Paix* (MJP) emerged south of the ceasefire line. These groups aligned themselves to the northern MPCI (the alliance came to be known as *the Forces Nouvelles.*) Both had close links with Taylor's Liberia and both vowed to avenge the death of Gueï (Staff Reporter, 2002). Taylor supported both groups, seeking revenge for his ally Gueï and access to San Pedro harbour for exporting natural resources. He also responded to Gbagbo's support of the anti-Taylor MODEL[6] rebels. After Taylor went into exile in 2003 many of his security forces and ex-NPFL combatants joined MPIGO and MJP rebel factions (Global Witness, 2003). On the other side ex-MODEL rebels can be found fighting alongside local Gbagbo-sponsored patriotic militia that emerged in response to MPIGO and MJP, the most important of which are the *Forces pour la Libération de Grand Ouest* (FLGO).

Western Ivory Coast has gradually taken over the Liberian war economy, including its practices and its players. In the southwest logging continues as if there never was a civil war. In the first half of 2004, 265,000 metric tons of timber were exported, compared to 270,660 tons during the first six months of 2002, before the civil war erupted (IRIN, 2004). Progovernment militia, chiefs, rebel warlords, timber companies and ordinary villagers reportedly indiscriminately fell giant hardwood trees. Other reports mention low technology gold mining, secretly using the forced labour of children, and violent clashes between different militia over villages in the vicinity of gold-rich rivers and mines in western Ivory Coast (ICG, 2004a).

Alongside the economy of plunder another conflict dynamic is becoming visible in the west of the country, that is, the discord between indigenous communities and migrant settlers. Although these conflicts revolve around material interests, notably land, the ethnic component has become a key feature. While national political struggles

[6] MODEL predominantly consist of Krahn, whose ethnic cousins are the Ivorian Guéré (or Wê), who in turn have prominent members within Gbagbo's FPI (ICG, 2003).

have triggered this type of conflict, its root causes must be traced through the history of coffee and cocoa farming in the region. Under colonial rule and the postcolonial Boingy government massive migration was encouraged, sometimes even forcefully, in order to supply labour to the then sparsely populated western regions.[7] Immigrants mainly included central Ivorian Baloé and northern Dioulas, but also Burkinabé from abroad. Gradually these immigrants acquired land from indigenous Bété and Wê farmers through the politics of *mise en valuer*,[8] which provided that any citizen was allowed to become the owner of the land that he or she worked, regardless sex, age or ethnic origin (Chauveau, 2000). Indigenous groups have ever since resented the fact that they did not equally profit from the economic development that grew out of their soil. In line with his antistranger rhetoric Gbagbo quickly adopted a programme of land reclamation, providing a sense of impunity to those who wished to chase immigrants off their land. Thousands of immigrant workers and farm-owners, mainly Burkinabé and Dioula, came under attack and now live in refugee camps. The great majority of the displaced are long-term residents of Côte d'Ivoire (Global IDP, 2005). Antiforeigner attacks in turn sparked retaliation attacks and rebel alignment on the side of the groups under siege. Most intense intercommunal fighting seems to take place within and south of the *zone de confiance* between the Burkinabé and Wê villagers, occasionally aided by respectively rebel and patriotic troops. Some reports speak of actual "ethnic cleansing" practices, taking the interethnic issue far beyond a material conflict over tenure.

The antiforeigner rhetoric that has been sweeping through the country for more than a decade has been adopted by local militiamen and security forces as a cover for violent plunder. Nationally it has given certain political leaders the opportunity to remove opponents in order to gain political and thereby economic power. In an International Crisis Group report (ICG, 2004b, p. 5) the question is posed "who would gain from peace and security in Ivory Coast other than ordinary citizens?" Following the report it will not be the soldiers controlling the various roadblocks in the country, or the rebels that trade arms and levy informal taxes in the north, and definitely not the political establishment that has secured rents from the cocoa and coffee agro-industry. The ethnic rhetoric that is used to mobilize political support for civil unrest, has sadly enough been localized to such an extent that it has infused violent cycles of rage and revenge between indigenous and long-time migrant communities. In these confrontations ethnicity has become a stake of its own, revealing the sinister interactions of greed and grievance motivated violence.

---

[7] Indigenous communities in the west include the Bété, Wê, Dida and Krou. They speak the languages of the Kru family and all but the Krou form important political support bases for Gbagbo's PDI. Indigenous Mande speakers include the Yacouba, Gouro and Gban. The latter groups are not aligned to the Gbagbo regime. They stand in regional competition with the former ethnic groups, occasionally leading to violent interethnic clashes. Many Yacouba have joined the MPIGO and MJP forces. General Gueï was Yacouba. Yacouba ethnic cousins in Liberia, the Krio, made up a large share of Taylor's NPFL (ICG, 2004a).

[8] This system was adopted by national governments in Guinea and Ivory Coast, creating mass movements of people to fertile parts of the country. In Liberia and Sierra Leone no such attempts to modernize land tenure have been undertaken, leaving allocation of land in the hands of village chiefs.

5 REINTERPRETING RESOURCE AVAILABILITY, YOUTH, AND ETHNICITY

The economic perspective to civil war, as laid out by Collier and associates (in Bergdal & Malone, 2000), has gained popularity over the last few years. It was the first good attempt to characterize the changing conditions of civil war in the post Cold War era; conditions that have most remarkably revealed themselves in West Africa's forest wars. The cases described in this chapter confirm that resources, including timber, played a major and sometimes critical role in West African rebellions. For example, the conflict in Sierra Leone may not have escalated as it did, despite disparate ideological positions, had Taylor not wanted a piece of valuable diamond mines. But this is not the same as arguing that rebel movements like the NPFL and the RUF were fighting only for diamonds. The case study analyses reveal that feelings of injustice, marginalization, and revenge play an important role at level of individual motives as well as collective objectives. While these grievance narratives can be explained as a mobilizing mechanism for rebel leaders and as an excuse for resource capture, they are alive and real for insurgents at all levels of the rebel organization. Therefore they cannot be denied as motives for rebellion. Greed and grievance are endlessly entangled in rebel motivations, also in countries where the violent struggle over natural resource wealth played an important role as the case studies reveal. The drastic conclusions by Collier and Hoeffler (1999, p. 15), that the demand for justice in rebels motivation is insignificant and that there is no evidence that the level of grievance is an important influence in civil war, are probably the result of their understanding of what are called proxies for greed and grievance based rebellions. These proxies are: Natural resource availability/dependence and (uneducated) youthfulness of the population as proxies of greed, and ethnic factionalism as a proxy of grievance. The meaning of the three central proxies needs to be reassessed considering the cases discussed in this chapter.

5.1 Availability of Lootable Resources

The availability of lootable resources, indicated by a country's reliance on primary commodity exports, features as the most important proxy of greed motivated rebellion. But the path of causation between resource abundance and conflict should not be perceived as a honey pot for rebels seeking loot and revenue. The degree on which states rely on these resources indirectly influences civil conflict because of its negative effect on economic performance and regime credibility. Using the phenomenon of Dutch Disease[9] Sachs and Warner (2001) explain how natural resource dependence slows down economic growth due to a decrease of investment in innovative economic sectors and an increase of an economy's vulnerability to price shocks. In addition, DeSoysa (2000) uses the term "rentier state" to indicate how resource dependent states are likely to be unresponsive and unaccountable to their constituencies as they do not

[9] The term 'Dutch Disease' refers to the negative economic effect of the overvaluation of national currency in the Netherlands, following the discovery of natural gas deposits in the North Sea in the late 1950s and early 1960s.

rely on taxes but safeguard revenues by strategic investment and alliance building. Eroding state functions in turn translate into decreasing investment in infrastructure, social service provision and law and order. Besides forming the object of looting, natural resource wealth, according to DeSoysa (2000, p. 125), "perpetuates bad governance, capricious political processes, underdevelopment and ultimately grievance." Natural resource dependence thus may equally link to greed as to grievance through its effect on national economic and political performance. The three countries discussed in this chapter were notorious for their patrimonial politics which evolved around natural resource sectors before the outbreak of civil war. Both Liberia and Ivory Coast witnessed negative growth rates of 2% before the civil wars broke out. In Sierra Leone, however, growth rates were still 2% in 1991, the year before the war.[10]

In addition to understanding the effects of natural resource dependence on economic and political performance, we should consider the physical terrain in which these resources are found. Rebels are surely drawn to natural resource hot spots to meet their economic needs and aspirations, but there may be other reasons why such areas are favored.

Several studies have suggested, although not convincingly proven,[11] that rebels tend to operate from peripheral mountainous and forest areas (Buhaug and Lujala, 2005; Collier and Hoeffler, 2001; Collier *et al.*, 2004; DeRouen and Sobek, 2004). The rebellions discussed in this chapter, apart from the MPCI in Northern Ivory Coast, all arose in remote cross-border forest regions. RUF strongholds were located in and around the three Gola Forest reserves, the single largest area of closed-canopy rain-forest remaining in Sierra Leone on the border with Liberia. The region was cut off from the rest of the country when an important railway line reaching the Gola region was closed in 1968 and German projects to tar roads to the area were never realized (Richards 1996:42–44). In this setting, beyond scrutiny and reach of central authorities, but with opportunities for illegal exploitation of natural resources and informal cross-border trading, the RUF found an excellent location to establish its base. In Liberia and Ivory Coast rebel factions found similar militarily strategic forest domains. In forested western Ivory Coast, Ivorian MPIGO rebels emerged at the exact same location (around Danané) as the Liberian NPFL did 13 years earlier. Danané is a small town, 20 km from the Liberian border, located on an important informal trading route between the two countries. During the Liberian civil war weapons passed through this town without any form of control. In 2002 MPIGO emerged around Danané where it met almost no resistance from the Ivorian army. Border guards immediately fled and gave themselves up to Liberian Anti Terrorist Units, Taylor's security forces (Global Witness, 2004a),

---

[10] Figures are taken from http://devdata.worldbank.org/query/default.htm.

[11] In their 2001 study Collier and Hoeffler conclude that civil war countries actually had slightly lower forest coverage (29%) than peaceful countries (31%). Also in terms of the duration of conflict, Collier *et al.* (2004) find that extensive forest cover is not significantly associated with longer wars. Buhaug and Lujala (2005) even find that densely forested countries tend to have shorter conflicts. In contrast DeRouen and Sobek's (2004) analysis of civil war outcomes shows that forest cover increases the likelihood of prolonged conflict.

indicating the lawlessness of the area. Also MODEL rebels have used the forest as hideout, their stronghold being Sapo National Park where some of them have remained until today, untouched by disarmament and repatriation efforts.

Remote cross-border forests in the region offered vulnerable rebel movements[12] a safe haven out of reach of government forces, but also a pool of resentful young people ready to join rebel movements. The reason for both can be traced to the history of state building in Africa. As a result of European preoccupation with ocean trade, the centres of powers of many African countries were drawn away from the traditional, interior regions to the coast (Herbst, 2000b, p. 17). Although some central axes were developed to facilitate the extraction of resources, the larger part of the countryside was not deemed economically profitable at the time and thus was left unexplored. These areas have remained devoid of effective administration, infrastructural development and economic investment. Where investment in natural resource sectors did take place, the uneven distribution of benefits, exclusion of local communities, and destruction of livelihood systems often fuelled resentment rather than compliance. In addition, economic crises since the 1980s and the decline of state armies after the end of the Cold War have reduced the ability of governments to exercise control over their territories. West Africa's remaining forest spaces contain a dangerous mixture of the aforementioned conflict conditions. The lootable resources in warfare often happened to be concealed in remote forest areas that additionally offered refuge and opportunities for rebel mobilization. Insurgents were thus not drawn to natural resource rich areas only because of their economic agendas.

5.2 The Youth Factor

The second and third indicator for greed based insurgence, used by Collier (2000, p. 94), are the percentage of youth in society and their lack of income-earning opportunities, measured by the level of education. Young men and particularly those with little prospects in life are most inclined to join rebellion. But what does this say about their motivation? One may equally reason that these factors evidence motivations of grievance. Richards (2003, p. 4) writes that "those who fight the war in Sierra Leone consider lack of education as their grievance!" While the economic dimensions in civil war are easy to be identified, symbolic dimensions that signify political and social complaints require deeper contextual knowledge. For example, it took some time for commentators to realize that cross-dressing and the wearing of wigs and comical masks by young fighters in Liberia should be interpreted as ritualized expressions of dissatisfaction with a society (for discussion see Ellis, 1999 & Richards, 2001). Is it mistaken that the economic opportunism of youth violence, because of its simplicity, comes to overshadow deeper resentments related to the lack of upward social mobility, intergenerational exploitation by elders, and uneven access to education.

---

[12] Liberian and Sierra Leonean rebel movements during their early incursions were very small. The NPFL of Liberia counted approximately a hundred men (Ellis 1999, p. 110) while the RUF in Sierra Leone started with only 35 (Abdullah & Muana, 1998, p. 147).

Equally important to mention is that motivations in and of themselves are for many young soldiers insignificant. Many young soldiers did not have a choice but were forced into war. This factor is hard to grasp in models of rebellion that are derived from market analogies, but must be considered particularly in the African context. Coercion can take many forms. In both Sierra Leone and Liberia, many children were abducted from villages to be socialized in military camps. The loyalty of these soldiers was guaranteed by sophisticated use of positive incentives, ranging from food and clothes, education, and life skills as well as negative incentives. In both Sierra Leone and Liberia negative incentives over-powered positive incentives as the war progressed (McIntyre & Addo, 2002). RUF runaways were maimed or killed. Even if they succeeded in escaping, some had the letters R, U, F scarred in their arms and thus risked being lynched by civil defence forces or executed by government troops (Richards, 2001, p. 75). In Liberia, boys were threatened with death either for themselves or their families if they refused to join the rebels (Twum-Danso, 2003). Besides direct coercion, family protection and peer pressure often played a key role. In militarized society where violence rules, joining an armed faction sometimes offers more security than staying on the sideline. Furthermore, tales of warrior cults communicated among peers link manhood to success in battle, providing strong cultural incentives to engage in warfare.

5.3 Ethnic Resentment

Considering economic motives to be key, Collier and Hoeffler (1999) deny that ethnic grievances importantly motivate conflict. According to their statistic analysis ethnic fractionalization (as a proxy to grievance motivated violence) does not make society more dangerous, it even makes it safer. They hold that in a fractionalized society rebel leaders have less chances of mass mobilization because the ethnic group they can draw from is smaller, reducing the chances of successful insurgence and ultimately civil war. Although the mobilizing potential of ethnicity is thus recognized, this understanding of the role of ethnicity in civil war is far too limited.

First of all, ethnicity in the case studies of Liberia and Ivory Coast was not used to create inward solidarity between rebels of one ethnic group. Instead we saw ethnic scapegoating of immigrant groups as a means to mobilize insurgency (in Liberia) and counterinsurgency (Ivory Coast) across ethnic divisions. Ethnic violence was manipulated by rebel leaders to serve strategic objectives, while in effect it also conferred impunity on followers who engaged in looting. But, as was shown, ethnic fault lines and oppositions hardened when the first shots fell. Incidental attacks would be considered a violation of an entire group's integrity by another group, and could therefore be retaliated indifferently. When such cycles of rage and revenge take precedence over looting opportunities, one may argue that ethnic grievances become an independent element in conflict.

In Sierra Leone we saw a different pattern. Here rebel leaders did not use existing ethnic animosities to mobilize people, but internally created new identities that would guarantee solidarity between recruits and captives. Fenced-off from the wider society an enclave culture developed that was based on equality and liberation ideologies in combination with bush knowledge. In such a sectarian enclave culture the increase in inward solidarity and identification was matched by the intensification of hostility towards the outside world, explaining some of the atrocious acts committed against rural neighboring villagers. Collier makes his speculations about the role of ethnicity using religious and ethnic categorizations from 1965, assuming that over a period of 30 years they probably have changed little (Collier, 2000, p. 95). To regard ethnicity in such narrow sense glosses over contemporary spiritual movements, sects, and subcultures that apply new identity constructs in defining their (violent) struggles.

The above analysis concurs that ethnicity is not a root cause of conflict. While ethnicity can turn into a separate motivation during later stages of conflict, it is principally deployed in an instrumental manner. However, to use ethnic fractionalization as a proxy to grievance motivated conflict is not correct. There is no reason to believe that grievance motivated violence is more likely to occur in heterogeneous societies than in homogeneous ones. First, the presence of a relatively well-off ethnic minority in society has mobilized grievances and contributed to humanitarian dramas throughout history. There is also no reason to believe that this is bound to happen more frequently in an ethnically fractionalized society. Second, the measure of static ethnic categories in a given country neglects upcoming insurgence groups that adopt new identity constructs to create inward solidarity and outward hostility. Again the degree of ethnic fractionalization in society says nothing about the likeliness of sectarian violence to emerge. Given that ethnic vocabularies can be applied equally in more and less fractionalized societies, one cannot use the degree of ethnic fractionalization to determine the extent to which ethnic grievances may or may not matter in civil war.

6 CONCLUSION AND POLICY IMPLICATIONS

The objective of Collier's analysis was to investigate the relative importance of greed and grievance as motivations for civil war. The findings of this study in summary were that the availability of natural resources and the proportion of uneducated youth in society positively correlate with the outbreak of civil war, while ethnic fractionalization does not. As much as the drivers of civil war are herewith identified, this kind of information does not permit any conclusions about the causes, understood as reasons or motives, of conflict. A critical investigation of the proxies applied in Collier's study based on case study analysis in West Africa challenges the bold statement that greed outperforms grievance as motive for civil war and calls for a critical assessment of policy recommendations deriving from the economic agenda literature.

The new perspective on economic agendas in civil war has generated a set of policy recommendations from several sources. DeSoysa (2000, p. 126) calls for measures to prevent illegitimate actors from gaining easy access to legitimate channels of international trade. This idea resonates in recent and current UN sanctions on timber and minerals in Liberia and the Democratic Republic of Congo and in broad based initiatives such as the Kimberly process that is meant to halt the trade of conflict diamonds (Herbst 2000). Over the longer term authors call for promoting economic growth and export diversification (Collier 2000, p. 107; DeSoysa, 2000, p. 127). These approaches certainly form an important part of reaching solutions to contemporary conflicts in West Africa. The focus on the economic dimensions of conflict, however, has two important shortcomings: First, it suggests that there are relatively "clean" ways to end conflict, where in fact there is a need for on-the-ground military intervention; secondly, perpetrators of violence are criminalized, in effect neglecting the desires and grievances that drove them into war.

Measures such as freezing bank accounts, sending out international arrest warrants and trade boycotts are in many cases unlikely to end fighting. Evidence from West Africa suggests that rebel leaders and soldiers have proven remarkably flexible and resourceful in dodging international measures and finding new areas of instability where they can pursue their warrior lifestyle (Herbst, 2000a, pp. 287–88). While international measures may take the zing out of the war economy, regional economic alternatives to war financing remain operative. This is due to weak or completely absent regional control on trade and trafficking, and enduring alliances between rebel factions and neighboring states. The robustness of West African rebellions calls for more direct and on the ground military presence of the international community aimed at ending violence, law enforcement and border control. Such foreign troop deployment can only be effective if units have sufficient fighting capacity and broad enough mandates to go after the key rogue players in violent conflicts. Politicians, however, are reluctant to enter into messy civil wars and involve their armies in actual combat. As a result when troop deployment is finally negotiated it is often too weak and has too little power to make a real difference. Recent experience teaches us that this was particularly true for intergovernmental missions under the flags of ECOMOG, the African Union and also the United Nations. Unilateral interventions by western powers in Africa have proven to be more effective. In 2000, British commandos were able practically overnight to bring an end to total anarchy in Freetown, Sierra Leone. The French *Operation Licorne* in 2002 effectively prevented a clash between northern militias and government troops in Ivory Coast, even though it created a military stalemate that continues until today. In order to effectively end African civil wars, intervention capacities of intergovernmental bodies should be strengthened so that they can send better equipped troops with broader mandates. If this cannot be achieved, one-to-one unilateral intervention by a western power is preferable to more politically correct but less effective joint operations.

The second shortcoming of the focus on the economic aspects of rebel motivations is that it denies the rebels' social and political agendas. Keen (2000, p. 39) rightfully argues that "though Collier is right to suggest that rebels may be reluctant to acknowledge the degree to which they are driven by greed, there are equal dangers in suggesting that the expression of grievances tells us nothing about their real motivation." Insisting on the claim that insurgents are mere bandits, a position to which the greed narrative is inclined, national governments and the international community play a very risky game. Denial of grievances in ending war and creating peace reinforces the sense of injustice felt by insurgents, risking the magnification of violence. Creating a peace that takes into account the grievances of all elements is a political endeavour that goes beyond a simple reconstruction of ordinary life to the way it was before the war started.

Grievance-targeted peace building begins with providing the incentives to entice insurgents out of war. The challenge here is to offer ways for young people to achieve respectability, security, employment and a sense of autonomy. These options may in some instances lead to restoring former rural communities, but often they do not. Instead alternative productive sectors must be identified. It must be recognized that during years of war, many young combatants have learned how to operate in natural resource sectors, e.g., tapping rubber trees, felling timber, digging for diamonds and transporting such goods. Indeed, they have, in Keen's (1998, p. 11) words, created "alternative systems of profit, power and security." Rather than trying to lead the ex-combatants back to a rural life they so eagerly wished to escape, the solution may be to bring the new activities started during warfare within the rule of law.

Sources of grievances originating from national politics are linked to neo-patrimonial networks of distribution and exploitation that marginalized large groups of uneducated rural dwellers as well as urban intellectuals. Soon after fighting has ended more viable, equal and transparent government institutions need to be created that can receive, redistribute, and invest taxes, aid money, and natural resource revenues. In particular investment in vulnerable remote rural regions, especially in the areas of education, infrastructure development and healthcare, is key not only to dispelling local peoples' feelings of abandonment, but also to reinstating administrative and military control.

A final, critical issue is the ethnic scapegoating as a means to mobilize (counter) insurgency. This problem deserves attention at intercommunity level, but also at national political levels where hostile ethnic narratives are often invented by politicians for strategic reasons. Fundamentally the problem comes down to the failure of the nation building processes in postcolonial Africa. Although nation building was eagerly initiated by most independence presidents, contemporary political leaders are not very keen on furthering this agenda (see also Chapter 6). Playing the ethnic card yields immediate political support and critical leverage in elections. Remedies to such practices include independent media, impartial judiciaries and broad-based political parties. Such democratic principles go hand in hand with rebel youths' calls for equal opportunities and transparent institutions in the run-up to civil conflict. Unfortunately these calls are often left unheeded, later to be silenced completely in the roar of warfare.

REFERENCES

Abdullah, I., & Muana, P. (1998). The revolutionary front of Sierra Leone. In C. Clapham (Ed.), *African Guerrillas* (pp. 173–178). London: James Currey.

Archibald, S., Bruce, B., Modad, W., Mulbah, E., Richards, P., Varpilah, T., & Vincent, J. (2005). *Community Cohesion in Liberia: A Post-war Rapid Social Assessment.* World Bank Social Development Papers, No. 21. Washington, DC: World Bank.

Atkinson, P. (1997). *The War Economy in Liberia: A Political Analysis.* http://www.odihpn.org/report.asp.

Baechler, G. (1999). Environmental degradation and violent conflict: Hypotheses, research agendas and theory building. In M. Suliman (Ed.), *Ecology, Politics and Violent Conflict* (pp. 76–112). New York: Zed Books.

Bergdal, M., & Malone, D. M. (Eds.) (2000). *Greed and Grievance: Economic Agendas in Civil War*. Boulder, CO: Lynne Rienner.

Buhaug, H., & Lujala, P. (2005). Accounting for scale: Measuring geography in quantitative studies of civil war. *Political Geography, 24*(4), 399–418.

Chauveau, J. P. (2000). Question Foncière et Construction Nationale en Côte d'Ivoire: Les Enjeux Silencieux d'un Coup d'État. *Politique Africaine, 17*, 94–125.

Clapham, C. (2003). *Sierra Leone: The Political Economy of Internal Conflict*. (EN) CRU Working Paper No. 20. The Hague: Clingendael Institute.

Collier, P. (2000). Doing well out of war. In M. Bergdal & D. M. Malone (Eds.), *Greed and Grievance: Economic Agendas in Civil War* (pp. 91–111). Boulder, CO: Lynne Rienner.

Collier, P., & Hoeffler, A. (1999). *Justice-Seeking and Loot-Seeking in Civil War*. Washington, DC: Mimeo. http://www.worldbank.org/research/conflict/papers/justic.htm.

Collier, P., & Hoeffler, A. (2001). *Greed and Grievance in Civil War*. World Bank Paper No. 31. Washington, DC: World Bank.

Collier, P., Hoeffler, A., & Sönderbom, M. (2004). On the duration of Civil War. *Journal of Peace Research, 41*(3), 253–273.

Crossette, B. (2000, July 6). Singling out Sierra Leone, UN Council Sets Gem Ban. *New York Times*.

Deacon, R., & Mueller, B. (2004). *Political Economy and Natural Resources*. Santa Barbara: University of California. http://www.repositories.cdlib.org.

DeRouen, K. R., & Sobek, D. (2004). The dynamics of Civil War duration and outcome. *Journal of Peace Research, 41*, 253–273.

DeSoysa, I. (2000). The resource curse: Are civil wars driven by rapacity or paucity. In M. Bergdal & D. M. Malone (Eds.), *Greed and Grievance: Economic Agendas in Civil Wars* (pp. 113–135). London: Lynne Rienner.

Duffield, M. (1998). Post-modern conflict: Warlords, post-adjustment states and private protection. *Civil Wars, 1*(1), 65–102.

Ellis, S. (1995). Liberia 1989–1994: A study of ethnic and spiritual violence. *African Affairs, 94*, 165–197.

Ellis, S. (1999). *The Mask of Anarchy: The Destruction of Liberia and the Religious Dimension of an African Civil War.* New York: New York University Press.

Fairhead, J. (2001). International dimensions of conflict over natural and environmental resources. In N. L. Peluso & M. Watts (Eds.), *Violent Environments* (pp. 213–236). Ithaca: Cornell University Press.

Global IDP Project. (2005). *Ethnic Clashes in Wild West, Displace Civilians, 2002–2005*. http://www.db.idpproject.org.

Global Witness. (2003). *The Usual Suspects: Liberia's Weapons and Mercenaries in Côte d'Ivoire and Sierra Leone.* London: Global Witness. http://www.globalwitness.org/reports/download.php/00068.pdf.

Global Witness. (2004a). *Dangerous Liaisons: The Continued Relationship Between Liberia's Natural Resource Industries, Arms Trafficking and Regional Insecurity.* London: Global Witness. http://www.globalwitness.org/reports/download.php/00212.pdf.

Global Witness. (2004b). *Same Old Story: A Background Study on Natural Resources in the Democratic Republic of Congo*. London: Global Witness.

Gongloe, T. S. (2004, Nov. 4). Send peace, not hate messages to your family and friends in Liberia: An appeal to Liberians in the diaspora. *The Perspective*.

Herbst, J. (2000a). Economic incentives, natural resources and conflict in Africa. *Journal of African Economies, 9*(3), 270–294.

Herbst, J. (2000b). *State and Power in Africa: Comparative Lessons in Authority and Control.* Princeton, NJ: Princeton University Press.

Homer-Dixon, T. F. (1994). Environmental scarcities and violent conflict: Evidence from cases. *International Security, 19*(1), 5–40.

Homer-Dixon, T. F. (1999). *Environment, Scarcity and Violence.* Princeton, NJ: Princeton University Press.

International Crisis Group (ICG). (2003). *The War is not yet Over.* Africa Report No. 72. Freetown/Brussels.

International Crisis Group (ICG). (2004a). *No Peace in Sight.* Africa Report No. 82. Paris/Dakar.

International Crisis Group (ICG). (2004b). *Liberia and Sierra Leone: Rebuilding Failed States.* Africa Report No. 87. Paris/Dakar.

Integrated Regional Information Networks (IRIN). (2004). *Côte d'Ivoire: Civil War Allows Rampant Illegal Logging*, Dec 23, http://www.irinnews.org/report.asp?ReportID=44801& Select Region=West_Africa.

Keen, D. (1998). *The Economic Functions of Violence in Civil Wars.* Oxford: Oxford University Press.

Keen, D. (2000). Incentives and disincentives for violence. In M. Berdal & D. M. Malone (Eds.), *Greed and Grievance: Economic Agendas in Civil Wars* (pp. 19–41). London: Lynne Rienner Publishers.

Keen, D. (2003). Greedy elites, dwindling resources, alienated youths: The anatomy of protracted violence in Sierra Leone. *Internationale Politik und Gesellschaft.* http://fesportal.fes.de/pls/portal30/docs/FOLDER/IPG/IPG2_2003/ARTKEEN.HTM

Le Billon, P. (2001). Angola's political economy of war: The role of oil and diamonds 1975–2000. *African Affairs, 100*, 55–80.

Libiszewski, S. (1992). What is an environmental conflict? Paper presented at the *First Coordination Meeting of the Environment and Conflicts Project (ENCOP)*, Bern/Zurich, April 30–May 1, 1992.

McIntyre, A., Aning, E.K., & Addo., P. N. N. (2002). Politics, war and youth culture in Sierra Leone: An alternative interpretation. *African Security Review, 11*(3).

Reno, W. (1997). Privatizing war in Sierra Leone. *Current History, 96*, 227–230.

Reno, W. (2000a). Shadow states and the political economy of civil wars. In M. Bergdal, & D. M. Malone (Eds.), *Greed and Grievance: Economic Agendas in Civil War* (pp. 43–67). Boulder, CO: Lynne Rienner.

Reno, W. (2000b). The real (war) economy of Angola. In J. Cilliers, & C. Dietrich (Eds.), *Angola's War Economy: The Role of Oil and Diamonds* (219–235). Pretoria: Institute for Security Studies (ISS).

Richards, P. (1996). *Fighting for the Rain Forest: War, Youth and Resources in Sierra Leone.* Oxford: James Currey.

Richards, P. (2000). Rainforest resource conflict: Would stakeholder analysis help pacify labouring casuals in Sierra Leone? In K. F. Wiersum (Ed.), *Tropical Forest Resource Dynamics and Conservation: From Local to Global Issues* (159–170). Wageningen: Wageningen University and Research Centre.

Richards, P. (2001). Are "forest" wars in Africa resource conflicts? The case of Sierra Leone. In N. L. Peluso, & M. Watts (Eds.), *Violent Environments* (65–82). Cornell: Cornell University Press.

Richards, P. (2003). Are there common causes to recent West African insurgencies? Economic opportunities and the war economy. In *Conference on Conflict and Development Policy in the Mano River Region and Cote d' Ivoire: The Regional Stakes for Stability and Reconstruction*, Paris, May 13–14, 2003.

Ross, M. (2003). *Natural Resources and Civil War: An Overview.* http://www.polisci.ucla.edu/faculty/ross/WBpaper.pdf.

Sachs, J., & Warner, A. M. (2001). The curse of natural resources. *European Economic Review, 45*(4–6), 827–838.

Staff Reporter. (2002). *Chaotic Conflict Deepens As Government Troops Fight To Recover Lost Territory in Ivory Coast.* http://www.allafrica.com.

Twum-Danso, A. (2003). *Child Combatants and Adult Wars in Africa.* Monograph No. 82. Institute for Security Studies (ISS). http://www.iss.co.za.

United Nations Security Council (UNSC). (2000). *Report of the UN Panel of Experts on Sierra Leone*. S/2000/1195. New York.

United Nations Security Council (UNSC). (2001). *Report of the Panel of Experts on the Illegal Exploitation of Natural Resources and Other Forms of Wealth of the Democratic Republic of the Congo.* S/2001/357. New York.

United Nations Security Council (UNSC). (2004). *Panel of Experts on Liberia Report.* S/2004/7522. New York.

Utas, M. (2003). *Sweet Battlefields: Youth and the Liberian Civil War.* Department of Anthropology, Uppsala University, Uppsala, Dissertation.

Vlassenroot, K. & Romkema, H. (2002). The emergence of a new order? Resources and war in Eastern Congo. *Journal of Humanitarian Assistance,* http://www.jha.ac/articles/a111.htm.

CHAPTER 4

NICARAGUA'S FRONTIER: THE BOSAWAS BIOSPHERE RESERVE

A. CARLA STAVER, WIL DE JONG, AND DAVID KAIMOWITZ

1 INTRODUCTION

In the search for solutions to the increase of violent conflict worldwide, the examination of correlates and causes has become increasingly important. A nontraditional approach to the issue of security should interest political scientists, economists, environmentalists, and policy-makers alike. This volume examines the strong geographic correlation between forests and incidences of violent conflict, which has stirred significant interest in the international development community (e.g., Blom *et al.*, 2000). Recent research estimates that roughly three-quarters of Asia's forests, two-thirds of Africa's forests, and one-third of Latin America's forests are located in countries that have experienced violent conflict in the past 20 years (Kaimowitz & Fauné, 2003). This geographic correlation suggests a relationship between violent conflicts and forests, although whether forests give rise to conflict or conflict decreases deforestation is unclear.

This chapter addresses both of these issues, with particular reference to Bosawas, on Nicaragua's border with Honduras, during the resolution of the famous war of the 1980s between the socialist Sandinistas and U.S.-backed Contras and the ensuing resource-driven conflict.

The consequences of extreme conflicts for the forest environment differ depending on the source of conflict. This chapter will argue that violent conflicts in forested areas frequently arise when groups vie for control over valuable resources, such as land or timber. However, in certain situations in which natural resources are not the roots of conflict, violence can actually protect forests by discouraging parties that might exploit resources legitimately (Dudley *et al.*, 2002). These protective aspects are often short lived and never outlast the conflict itself (McNeely, 2003). More frequently, modern wars have resulted in serious environmental degradation, through direct exploitation (e.g., Global Witness, 1995; see also Chapter 3) and indirect secondary effects (Dudley *et al.*, 2002).

W. de Jong, et al. (eds.), Extreme Conflict and Tropical Forests, 57–74.

Collier and Hoeffler (2001) suggests that modern conflicts arise where stewardship of valuable resources is contested, rather than where resources are scarce (e.g., Homer-Dixon, 1999) or for ideological reasons, as is traditionally suggested. This hypothesis explains violent conflicts in which groups fight for control of valuable timber (i.e., conflict timber, Global Witness, 1995; Global Witness, 2002b; see also Chapter 7) minerals (e.g., conflict diamonds, or oil, Global Witness, 1998; Global Witness, 1999; Global Witness, 2002a), or other resources, which can often be found in forests.

Conflicts over rich natural resources are aggravated by the limited government presence, the institutional vacuum, and the lack of law enforcement in remote forested areas that result from their inaccessibility. Lawlessness makes conflict all the more likely because it renders resources in remote areas vulnerable to exploitation by the first or strongest taker. This situation has been documented repeatedly where road construction provides access to previously isolated forests—where a forest frontier forms. New road construction in the Brazilian Amazon under the *Avança Brasil* development plan, for instance, promises to result in large scale, indiscriminate environmental degradation and associated social conflicts between disenfranchized smallholders, large soybean producers, cattle ranchers, and other rural stakeholders (Fearnside, 2002; Nepstad *et al.*, 2002). In 2001 alone, five rural leaders were assassinated as conflict among competing interests in the area escalated (Nepstad *et al.*, 2002). Conflict in turn encourages exploitative instead of appropriative land use; long term ecological integrated management is neither valuable nor viable where land tenure is uncertain (Hotte, 2001). Thus, understanding modern frontiers is important not only for preventing and resolving violent conflicts but also to prevent and manage often irreversible environmental destruction that occurs at the frontier.

With a case study from Nicaragua, this chapter describes conflict arising from the formation of a frontier. The end of the 1980s Cold War-era war between the socialist Nicaraguan army and the U.S. backed Contras made large areas of Nicaragua's Bosawas region (Figure 4.1) dangerous and, in ecological terms, protected from unsustainable exploitation. The resolution of the civil war made the Bosawas region safe and available for settlement, leading to the formation of a frontier. First, this chapter will discuss the frontier's most important conflict-related characteristic: The absence of effective government presence. This institutional vacuum is conducive to the escalation of conflicts of interest into violence. Belated attempts on the part of governments to establish institutions can actually exacerbate conflicts, because settlements are often not appropriate for building lasting peace. Settlement of the Bosawas region in the early 1990s gave rise to conflicts over land tenure and land availability, which were escalated in part by an inappropriate government response. Subsequent mediation in the region shows how careful planning of conflict resolution, including considerations of local needs and dissemination of power to local peoples, can contribute to constructive and lasting peace in conflict-prone areas. With additional examples, we will conclude with recommendations on conservation and management measures to de-escalate conflict and mitigate its effects on the environment.

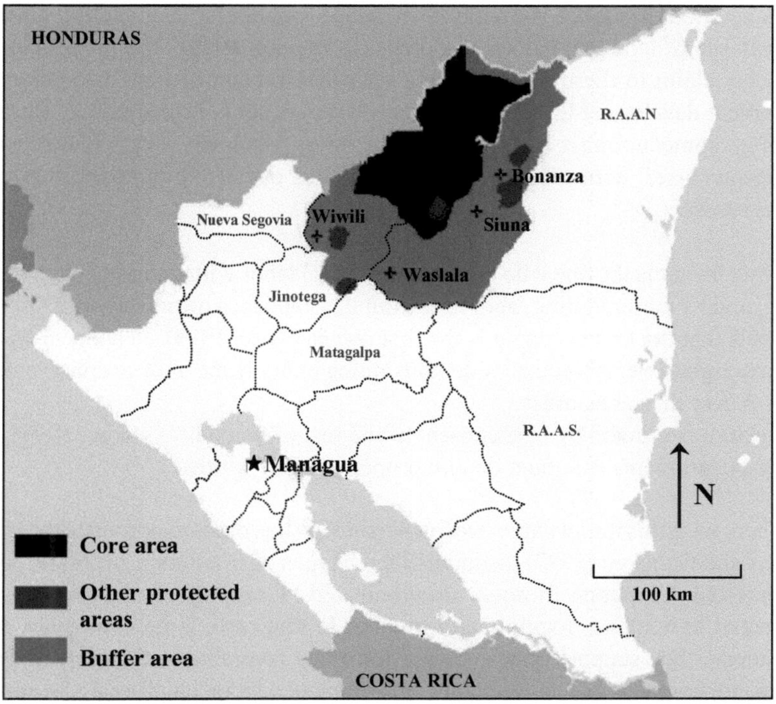

Figure 4.1. *Nicaragua and the Bosawas Biosphere Reserve (adapted from INETER 1999; MARENA 2001).*

2 THE FRONTIER

The nature of the frontier is such that violent conflict often arises and that unwonted environmental degradation occurs. An understanding of the factors that contribute to violent conflict and environmental degradation at the frontier may help policy makers avoid these outcomes. So what makes a frontier?

Our understanding of the frontier has improved somewhat since Turner (1893) described it as "the meeting point between savagery and civilization" in the work that lay the foundations for frontier studies. Nonetheless, the phrase communicates an essential point: Frontiers form where people rapidly migrate to previously sparsely populated areas, creating a meeting point between the "developed" and "undeveloped" world. An alternative defines the beginning of the frontier as the point in settlement of an area at which a market economy becomes possible (Schneider, 1995). These fundamental definitions are important (see Bilsborrow, 2002; Carr, 2004), but do not necessarily describe what makes the frontier conducive to violence or environmental degradation.

Perhaps the key consideration in the emergence of violent conflict at the frontier is the determination of property rights. Frontier economies are unique in that they are characterized by low—if rapidly increasing—population and capital density and an abundance of land and natural resources (Schneider, 1995). The process by which settlers lay claim to these resources is a vital determinant of how the sociopolitical environment develops at the frontier. "How do things get to be owned? . . . Back at the beginning, someone must have acquired the thing, whatever it is, without buying it from anyone else," writes Rose (1994). Essentially, ownership is established through two mechanisms:

- The first owner is the one who "mixes his [or her] labor with the previously unowned thing, and by commingling labor to the thing, establishes ownership to it." A variant suggests that the first occupant is the first owner (Rose, 1994). In short, people earn de facto rights over resources by demonstrating publicly that they exercise significant control over those resources.
- Ownership is granted by "the consent of the rest of humanity" (Rose, 1994). Effectively, governments can grant de jure property rights or titles.

Who, then, wins this initial battle over ownership? Who comes to dominate the frontier? Browder and Geoffrey (1997) distinguish three types of frontiers: Corporate, populist, and contested. Corporate frontiers are dominated by capitalized enterprises, such as commercial agriculture, logging, and mining. In this case, large companies and the governments they support impose ownership rules favorable to themselves. Populist frontiers are composed of autonomous, self-employed labor, small-scale farming, logging, and mining operations, and local merchants. Most frontiers include elements of each; Browder and Geoffrey (1997) call these contested frontiers, where neither group successfully dominates. Conflicts over ownership arise when property rights are determined by de facto occupancy or dominance. Thus, the benefits of de jure property rights in a frontier region are clear, although the methods and institutions that determine who receives property are especially important for sustainable and equitable development (see Schneider, 1995).

However, frontier regions are characterized by poorly developed institutions, in large part because developing government institutions is expensive and possibly even cost-ineffective during the early phases of frontier formation (Alston & Mueller, 2003; Schneider, 1995). In most cases, a frontier's first colonists are smallholders simply because this process is economically efficient and probable (Schneider, 1995). However, economists tend to suggest, even while propounding the benefits of equitable and sustainable development and the necessity of secure land tenure for small farmers, that imposing external government on a developing frontier tends to detract from the profitability and efficiency of the frontier. In a World Bank environment paper, Schneider argues that an "emerging" government—one that arises in situ in response to the needs of frontier settlers—is more desirable than an "imposed" government—one that governs from beyond the frontier (1995). Likewise, Alston and Mueller suggest that "imposed" government is not appropriate simply because it is costly (2003). In Brazil, for instance, capacity building for municipal and regional governments has aided in decentralization of development planning and conservation, with some success (Nepstad *et al.*, 2002).

However, a total absence of an external government that reinforces local government and the property rights of smallholders often means that they lose their rights to land. As the frontier develops, the value of land and resources increases, resulting in a second wave of wealthier settlers who are unwilling to take the risks associated with the frontier at its inception (Schneider, 1995). If the original settlers have legal and well-enforced rights to their land, they can retain it or, at the very least, sell their land. However, settlers are less likely to secure titles to the land they occupy when their land is more remote, as occurs on a frontier. They are often ejected by wealthier claimants, who wield more political influence (Hotte, 2001).

Where centralized land granting systems do exist, they sometimes contradict systems established in situ. This has been the case with many emerging frontiers in Amazonia. Brazil's National Institute of Colonization and Agrarian Reform (INCRA), established in 1970, is subject to pressure from interest groups, which tend to represent wealthier stakeholders rather than the original settlers of the frontier (Hotte, 2001), who tend to be poor (Schneider, 1995). The *Bico de Papagaio* area of Tocatins in Brazil, for instance, is famous for its bloody land conflicts between small farmers and large-scale *grileiros*, or land grabbers (Fearnside, 2001). Paraguay's land titling law changed from a system which favored small farmers during the 1960s to a system that favored the wealthier elite during the 1970s and 1980s; conflicts arose among claimants that ultimately resulted in widespread violence and eventually, when questions over human and civil right emerged, a coup d'etat in 1989 that overthrew Alfredo Stroessner Matiauda (Nagel, 1999). In Mozambique during the late 1980s, conflicts between subsistence and commercial farmers arose simply because neither held recognized land titles (Hotte, 2001).

In special cases, illegal activities can flourish under the total absence of centralized government and law enforcement. Indeed, local government does emerge, but these well-established groups do not contribute to the peace and stability of the frontier. For instance, drug cartels are well established in frontiers around the world, including the Chapare region in Bolivia, the Huallaga Valley in Peru, the Colombian departments of Caqueta and Guaviare, and the Golden Triangle in Burma, Laos, and Thailand (Koutouzis & Perez, 1996), where they benefit from the lack of governance at the frontier. The illegal production of conflict timber, conflict diamonds, or oil (see Global Witness, 1995; Global Witness, 1998; Global Witness, 1999; Global Witness, 2002a; Global Witness, 2002b; see also Price et al., this volume), where powerful, nondemocratic and nongovernmental groups control an area, are equally good examples. In these situations, local peoples and entire nations alike are the losers, illustrating the costs of a truly hands-off approach to frontier formation.

Another major issue associated with the emergence of a frontier is the severe environmental degradation that frequently occurs there. To some extent, resource exploitation and settlement which drive the expansion of the frontier make ecological degradation inevitable (e.g., Bilsborrow, 2002; Carr, 2004; Whitmore, 1999), but lawlessness and especially insecure land tenure seriously aggravate unsustainable use of forests and resources. Again and again, insecure land tenure has been shown to result in high levels of deforestation, unsustainable resource use, and low levels of investment and management (e.g., Chomitz & Gray 1996; Godoy *et al.*, 1998; Hotte, 2001; Schneider, 1995). Conflict itself can foster groups that seek to exploit resources unsustainably and exclude parties interested in instituting sustainable and equitable development and conservation schemes. Conversely, good governance at the Amazonian frontier in Brazil, for instance, and careful implementation of road construction projects with a view to promoting management and sustainable use may help stop the rampant degradation that has accompanied historical "business-as-usual" frontier expansion in Brazil (Fearnside *et al.*, 2005; Nepstad *et al.*, 2002).

Nicaragua's north-central region (the Bosawas Biosphere Reserve and the surrounding areas in Matagalpa, Jinotega, and the North Atlantic Autonomous Region) provides an excellent case study of the close relationship between forests and extreme conflict under frontier conditions. Following the civil war that ravaged the region in the 1980s, various rebel groups were able to thrive under the frontier conditions that existed there in the 1990s. Overlapping claims on frontier forest resources in the context of weak government presence fueled nearly continuous conflict in this region.

3 HISTORY OF WAR IN THE 1980S

The west of Nicaragua is characterized by high population density relative to the much larger eastern departments, which are inhabited mostly by *mestizo* and indigenous peoples. British and North American culture has had a deeper effect on the eastern region, including the area of the Bosawas Biosphere Reserve (see Figure 4.1), than the Spanish influence that shaped the rest of Nicaragua. One of the largest remaining areas of tropical forests in Central America is found in the North and South Atlantic Autonomous Regions (RAAN and RAAS) and Rio San Juan department on the Atlantic Coast. Indeed, any significant remaining forest in Nicaragua lies in this eastern region; the west has been almost entirely deforested.

The departments of Zelaya (which would become the RAAN and RAAS in the late 1980s), Matagalpa, Jinotega, and Nueva Segovia were the most distant from the rest of the country and from national government control. It should, therefore, come as little surprise that the growth of the resistance to the Sandinista government occurred in these very departments, following the Sandinista overthrow of dictator Anastasio Somoza in 1979 whose family ruled Nicaragua from 1936 to 1979 (Merrill, 1993). Two major groups formed: The Nicaraguan Democratic Force (FDN) and the MISURA indigenous group. The region's proximity to Honduras allowed these groups to establish supply networks easily, and forests provided ideal cover for rebel groups. Lack of a Sandinista government presence in the area meant that rebel groups could operate with more freedom than elsewhere.

3.1 Fuerza Democratica Nicaraguense—The Contras

Members of Somoza's National Guard formed the FDN, popularly known as the Contras, in 1981. Although small farmers and peasants in the north of the country originally supported the Sandinista movement, they soon became disillusioned with the party's social and agricultural programs, which created farm cooperatives out of land holdings confiscated from Somozistas and others. Many small farmers saw no benefits of socialization and accordingly began to support the Contras. Those who did not choose to join the Contras were forced to join Sandinista armies and militias to defend themselves.

Nationally, the Contras formed the other half of the ideological dichotomy in Nicaragua during the 1980s in opposition to the Sandinistas' socialism. It was a manifestation of the Cold War polarity: The Soviet Union supported the Sandinista government, while the United States backed the Contras. As was often the case during the Cold War, the two sides, motivated by ideological concerns, took advantage of local interests and conflicts to boost support for their cause. On the local level, however, competition over natural resources rather than ideology often precipitated violent conflict, much as it does today.

The FDN never had the military prowess to present a real threat to the capital, Managua, or any other major cities, but they did manage to control the area in and around what would become the Bosawas National Reserve in 1991 and later the Bosawas Biosphere Reserve. In that area, the aid they received from the Reagan administration helped them to inflict extensive economic and human damage (Merrill, 1993). However when the Iran-Contra affair, involving the illegal sale of arms to Iran to raise money to fund the Contras war was leaked to the American public, the US administration stopped aid in 1987. The FDN movement quickly fell apart.

3.2 MISURA—The Miskitos

Farther east, the Miskito Indians of the Atlantic Coast also became dissatisfied with their role and rights within the new government (Merrill, 1993). The region had historically enjoyed very little government interference. The native inhabitants, mostly Miskito and Mayagna Indians, initially welcomed improved education, health care, and access to credit that the new government introduced into the region but came to resent the interference that this necessitated and the wholly different culture that accompanied it: Spanish replaced English, Miskito, and Mayagna; the Catholic replaced the Moravian church. Local peoples began to demand autonomy soon after the foundation of the Sandinista government.

Indigenous and government individuals in the area formed MISURASATA (**Mi**skito, **Su**mo, **Ra**ma, **Sa**ndinista Asla **Ta**kanka [United]) in 1981 in an attempt to promote community cooperation, but relations between the group and the Sandinista government soon deteriorated to the point that the Sandinistas started arresting indigenous leaders of MISURASATA. Several Miskito leaders fled to Honduras, where they formed MISURA (**Mi**skito, **Su**mo, **Ra**ma). The Reagan administration provided similar support as the Contras received, including arms, money, and training to combat the Sandinistas (Kaimowitz & Fauné, 2003). Thus, the Contras and MISURA occupied similar positions on the political spectrum, although they at no time coordinated their efforts. In direct combat with the Nicaraguan government, the MISURA were

Table 4.1. *Acronyms used in this chapter.*

Bosawas B.R.	Bocay River, the Saslaya Mountain, and the Waspuk River Biosphere Reserve
FDN; Contras	Nicaraguan Democratic Force
FN 3-80	Northern Front 3-80
FUAC	Andrés Castro United Front
MARENA	Ministry of Environment and Natural Resources of Nicaragua
MISURASATA	Miskito, Sumo, Rama, Sandinista United
MISURA	Miskito, Sumo, Rama
PAF-CA	Tropical Forest Action Plan for Central America
RAAN	North Atlantic Autonomous Region
RAAS	South Atlantic Autonomous Region
RN	Nicaraguan Resistance
YAMATA	Descendants of the Mother Earth

generally even less successful than the Contras, and eventually most of the Miskitos and Mayagnas either fled to Honduras or were forcibly relocated to resettlement camps.

American support for the MISURA insurgency had nothing to do with the land and resources in the region, but it certainly capitalized on the demands for land tenure and stewardship by local and indigenous groups. MISURA was already negotiating for autonomy, which it eventually received under Sandinista governance. The Sandinista government agreed to negotiations with MISURA and responded to the complaints of the resident population only after a continued government military presence became clearly costly and ineffective. In 1987, the negotiation culminated in two autonomous regions: The RAAN near Bosawas and the RAAS. Although autonomy granted the major demands of MISURA, various splinter groups formed, of which the largest was YATAMA, an acronym for a Miskito phrase meaning "descendants of the mother earth" (Kaimowitz & Fauné, 2003) (see Table 4.1).

3.3 A Resolution: Resettlement

In response to international and internal economic pressure in the late 1980s, the Sandinistas moved up scheduled general elections to February 1990 (Merrill, 1993). Violeta Barrios de Chamorro, the National Opposition Union (UNO) candidate and editor of *La Prensa*, the major Nicaraguan newspaper, won the election, and shortly thereafter signed peace agreements with the FDN and YATAMA. Under these agreements, ex-combatants and their families were given land in unclaimed forested areas near what would become Bosawas Biosphere Reserve. Additionally, many thousands of decommissioned Sandinista soldiers required land and resources; officers were also settled in the area, although soldiers were for the most part left to fend for themselves (Kaimowitz & Fauné, 2003).

The process of resettlement set the stage for conflict in the decade to come. Throughout the 1980s, the region had experienced some conflict over land and resources, which led to the supposedly ideologically polarized conflict in the region during the revolution. However, the frontier characteristics of the region during this decade were small-scale.

The resettlement process created a veritable frontier. Not only did it open up a vast area of natural resources, but it did so in a way that inevitably created intense conflicts between resettled insurgents and indigenous residents, and even among the settlers themselves. Moreover, many decommissioned soldiers from the Sandinista military still had no land or livelihood. Groups that had fought each other for a decade were awarded conflicting land claims as a cure-all solution.

The Nicaraguan government had no local mechanism for responding to the conflict that arose. The area had been almost completely abandoned because the central government saw nothing of value there. However, as soon as the government decided upon resettlement as a key element in its policy for the promotion of peace, the land and forest in this region became attractive and increased in value. As people migrated to the area, a frontier developed, conflicts emerged, and the need for effective government institutions arose. The national government could not develop sufficient presence quickly enough to settle disputes effectively or to prevent further conflict.

4 THE BOSAWAS BIOSPHERE RESERVE

In 1991, the Nicaraguan government added its own additional claim to the land in the area when President Chamorro created the Bosawas National Reserve of Natural Resources, officially as a part of the Tropical Forest Action Plan for Central America in an attempt to contribute to the improvement of local ecological conditions. Politically, this may have been a move to garner support from the international community. Regardless of its direct motivations, however, the creation of the park reflected the value placed on biodiversity by both international and national parties, who viewed it as a resource more valuable than those that could be directly exploited. The reserve includes approximately 8,000 km^2 of mostly primary forest. It is the home of 2,500 tree species, 12 endangered animal species, and 18 others animal species extinct in the rest of Central America ((Kaimowitz & Fauné, 2003; WRM, 2000). The landscape ranges from rolling hills to steep slopes, from which flow a network of rivers, streams, and creeks, which have historically provided the only means of transportation in the area. The name Bosawas is an acronym formed from the first letters of the **Bo**cay River, the **Sa**slaya Mountain, and the **Was**puk River.

The Nicaraguan Minister of the Environment at that time understood that with the end of the war, the region would come under increasing pressure from mining and timber companies and from agricultural expansion, as is often the case in forested areas immediately following times of conflict (e.g., McNeely, 2003; Ch 10), but especially so because the region was the active resettlement site for demobilized fighters from the FDN, MISURA, and Sandinista army. Nonetheless, the Bosawas National Reserve began as a "paper" park. Because of lack of funding the government was unable to engage sufficient management staff for the park and had no existing law enforcement network in the region. The Minister negotiated for financial support from the German Agency for Technical Cooperation (GTZ), the United States' Agency for International Development (USAID), and the Nature Conservancy (TNC), which were available starting in 1994. Insufficient park management and law enforcement allowed the armed groups of the 1980s to reestablish themselves in the area.

Though the area was at first sparsely populated, its inhabitants increased by almost 50% after the end of the war in 1990. These inhabitants were often unaware that they lived in a designated protected area. In 1996, the Ministry of Environment and Natural Resources divided the Bosawas National Reserve into two regions: The core area of 3,300 km^2 where no human activities were to be allowed and a buffer zone with certain specified land and forest-use restrictions (Kaimowitz & Fauné, 2003). In 1997, UNESCO declared Bosawas a Biosphere Reserve. No additional conservation activities or funds accompanied this declaration.

5 THE THREE GROUPS IN CONFLICT

Conflicts over land and resources in the Bosawas Biosphere Reserve area inevitably developed. Active resettlement during the post-war period had created a true frontier in an area that had previously been regarded as invaluable and therefore lacked institutional development. The historic conflict in the area aggravated the climate of lawlessness. The creation of the Bosawas Biosphere Reserve from 1991 to 1997 added another competing claim in the area, one which was impossible to enforce. Funding for park management finally arrived in 1994, but enforcement of the conservation function of the park was not received well by settlers, who felt they were being deprived resources that were rightfully theirs. Ex-MISURA, ex-Contra, and ex-Sandinista settlers reacted by forming new rebel groups.

During the 1980s, American interests had used local grievances and demands for autonomy and land to fuel a national-scale uprising, especially among the FDN, to destabilize the Sandinista government. The groups formed in the early 1990s received no such backing and were consequently not complicated by ideological rhetoric. Instead, their violent insurgency was aimed largely at procuring immediate improvements in their material situation, and they negotiated accordingly. The national government too was no longer a party to Cold War era ideological conflict, and its ascendancy was never threatened. Rather, it negotiated to end violence in the Bosawas region through a series of concessions to insurgents.

5.1 Northern Front (FN) 3-80

The Northern Front 3-80 (FN 3-80) was a descendant of the Nicaraguan Resistance (RN), an alliance between the FDN and another resistance group, the Revolutionary Democratic Alliance. The RN had disarmed in 1990 in return for land for retiring fighters near Siuna and Waslala, towns that were designated "development poles." However, the Nicaraguan government and foreign development assistance agencies soon discovered that they could not meet the demands of the RN fighters for land, credit, and social services. Whatever the real reasons for this lack of funds, ex-Contras thought the government was favoring the demands of opposing groups. They observed with mounting rancor the distribution of land to ex-Sandinistas, the willingness of the government to allow Sandinista soldiers to keep the land they had confiscated during the war, and the creation of the Bosawas National Reserve.

Ex-Contras formed several groups to represent their interests, of which FN 3-80 soon became dominant. Among other things, the group demanded the resignation of suspected Sandinista sympathizers in the President's cabinet, including Minister of Defense Humberto Ortega and Minister of the Presidency Antonio Lacayo, the President's son-in-law. They also demanded that the government guarantee the safety of former RN fighters.

At various times between 1992 and 1997, the government gave FN 3-80 virtual control over parts of the Cua-Bocay, Waslala, and Wiwili municipalities, where the group imposed vigilante-style order, killing thieves, cattle rustlers, and rapists, as well as some hundreds of Sandinista sympathizers, and in effect filling the existing law enforcement vacuum. Various foreign and national agencies confirmed the dominant position of the FN 3-80 by consulting them on matters of development, agriculture and land titling.

After negotiations failed in 1993—FN 3-80 first took Nicaraguan hostages and then captured two French military attaches—the Nicaraguan army and FN 3-80 agreed to a ceasefire in 1994. During the subsequent negotiations, both parties signed a disarmament agreement in which the government promised to provide land, credit, and technical and medical assistance to the fighters and agreed to pay for all automatic weapons handed over by FN 3-80.

Many FN 3-80 members did disarm and move to security zones. However, several commanders, especially one Sergio Palacios, refused to do so because they felt that promises of land, credit, and assistance were insufficient (Cuadra & Saldomando, 1998). When the army began to exercise what little control it had in the region, negotiations began to succeed. The army assassinated Palacios in 1996. In 1997, the government pledged land, food, clothing, seeds, housing materials, and social services for former FN 3-80 members in return for disarmament, in addition to amnesty and security guarantees (Immigration and Refugee Board of Canada, 1997). As of 1998, the FN 3-80 formally disarmed, although some of its members formed criminal bands, kidnapping and assaulting people and stealing cattle. The group as such is no longer active.

The FN 3-80 certainly claimed to fight for RN ideals, making some half-hearted national-level political demands. Fundamentally, however, their cause was not political. Ex-RN fighters re-armed in the early 1990s because they resented the allocation of limited resources to their perceived rivals, ex-Sandinistas. They settled for land in the Bosawas region and other immediate material gains. Meanwhile, the government, because its influence was so limited, appeared to have no immediate alternative solution to the FN 3-80's violence other than further promises of land and services.

5.2 Andrés Castro United Front (FUAC)

Following the Sandinista electoral defeat in 1990, thousands of Sandinistas in the government army were demobilized, and became as frustrated and disgruntled as the FN 3-80 Contras. While the Contras complained that Sandinistas were allowed to keep much of the property they had confiscated from large landowners during the war, ex-Sandinista soldiers protested that much of their property had been reinstated to its prewar pro-Somoza owners. They also resented that those same high-ranking Sandinistas in the military against whom the FN 3-80 protested were turning their backs on their old supporters. Mostly, these soldiers found, however, that as soon as the Sandinista government was removed from power, they were landless, with no credit, training, employment or social services. Several thousand took up arms in protest in 1992 under the leadership of former military officials, including Edmundo García and Gustavo Navarro. They founded the Andrés Castro United Front (FUAC), named after the Nicaraguan populist hero Andrés Castro who fought against William Walker in the 1950s.

Farmers who had belonged to agricultural cooperatives during the Sandinista years were disappointed that the government seemed unwilling to provide them with land titles and to continue to support them with credit, health care, education, and infrastructure maintenance. The FUAC capitalized on this disappointment to build a large base of support in the area, which they reinforced by killing cattle-rustlers and other suspected criminals. The FUAC did not publicly declare its existence until July 1996. Although it did defend these farmers and criticize President Chamorro's policy and that of her successor, Arnoldo Alemán of similar political orientation, the FUAC ultimately campaigned for little more than resource access for its members.

In 1997, the FUAC and Alemán signed a peace agreement in which the FUAC would first be confined to peace enclaves, a concession of control over restricted areas, two of which bordered Bosawas, and eventually would disarm. In return, the government agreed to provide land, health care, and scholarships to the soldiers, as well as six months worth of food provisions. The government also agreed to allow the FUAC to establish a foundation to raise international funding for housing, credit, infrastructure and professional training.

As with the disarmament of the FN 3-80, the disbanding FUAC left behind splinter groups, the largest of which was the Revolutionary Armed Forces, allegedly led by García and Navarro, who had founded the FUAC. However, even the assassinations of these two leaders in 2000 did not stop the violence of armed bands that continued to operate in the area. Newspaper articles reporting hits by government forces against FUAC militants continued into 2002. The similarities between the FUAC and the FN 3-80, though supposedly politically opposed, are striking. Although they campaigned for ideological ends, the FUAC settled for—and perhaps only really ever wanted—land and material assistance. Here again, because of the limited government presence characteristic of a frontier, granting these concessions was the national government's most efficient means to at least a temporary peace.

5.3 The YATAMA

The "Descendants of Mother Earth" (YATAMA, for its name in Miskito) first appeared on the political scene in the Bosawas area around 1984 when the Sandinista government first began negotiations with MISURA fighters. It comprised a diverse council of indigenous leaders that represented a unified indigenous cause in the Bosawas/RAAN area. As MISURA declined, the YATAMA became more prominent and took over MISURA's role as sometimes violent advocates of indigenous territorial and civil rights in Bosawas.

While the FN 3-80 and the FUAC were similar in many respects, the YATAMA were different. First, the YATAMA had no central leader. They were instead an assembly of indigenous military commanders and their followers; many were embittered by their experiences under a central command during the 1980s. Organizational difficulties meant that for the most part, the YATAMA engaged in no coordinated attacks. Rather, they were involved from 1992 to 1995 in sporadic acts of violence.

The central tenet of the YATAMA alliance was the fight for autonomy and demarcation of indigenous lands. However, specifics were unclear; the autonomy law of 1987 certainly did not answer their needs, undoubtedly because it was not actually ever implemented, and most did not request community land titles. The YATAMA continued to demand recognition of one large indigenous territory, as opposed to small community ones, and requested compensation for logging, mining, and fishing concessions in their territory. However, despite ongoing demands for autonomy and self-determination, the YATAMA repeatedly negotiated for the same short-term benefits as the FUAC and the FN 3-80: Housing, credit, food, and other services. Periodically the government made small concessions, but it rarely delivered on promises. At one point, the government asked a US backed forestry project to hire and train demobilized fighters as park rangers, but this, as well as some of their other agreements, was a negligible provision.

In 1997, the Ninth Assembly of Indigenous People and Ethnic Communities voted to legitimize Miskito claims for autonomy and demarcation of indigenous territories. A year later, demobilized YATAMA fighters re-armed and took control of part of the RAAN (Flores, 1998). Negotiations with the central government followed the prompt withdrawal of the Nicaraguan army, in which the Nicaraguan government agreed to help with the demarcation of indigenous territories and to provide immediate assistance with housing, credit, and land. In early 1999, the YATAMA again officially disarmed (Lopez, 1999).

However, like the FN 3-80 and the FUAC, the YATAMA were dissatisfied with the government's fulfillment of its promises and continued to threaten to again take up arms. This was perhaps inevitable. Traditional indigenous territories, for which indigenous peoples did not officially have titles or autonomy, were reallocated to decommissioned soldiers from the civil war of the 1980s, creating land conflicts among the many resulting claimants. In addition, the Bosawas Biosphere Reserve presented another claim to land and resources in the area, rather than a possible solution to the problem.

However, unlike the FN 3-80, which never gained popularity, and FUAC, which never earned national legitimacy because of their violent and often quasi-terrorist tactics, the YATAMA eventually chose to participate in legitimate government. In the 2002 RAAN elections, the YATAMA came in third place, behind the Liberal Party (PLC) and Sandinistas (FSLN). They formed a coalition with the FSLN to become a major political force in the department (NicaNet, 2002). In 2003, the Nicaraguan

National Assembly unanimously approved codification for the 1987 Autonomy Law, giving regional leadership in the RAAN and RAAS a measure of stewardship over institutional structure and policy and over natural resource management, in cooperation with national institutions (NicaNet, 2003).

6 DISCUSSION

Continuing conflict during the 1990s in the Bosawas Biosphere Reserve region directly arose from the resettlement of demobilized soldiers and decommissioned officers from the Nicaraguan Resistance (RN) and Sandinista armies. Road construction alone might have created an environment in which these settlers and the indigenous and *mestizo* peoples who already lived in the area clashed over control of land and resources. Active settlement plans on the part of the central Nicaraguan government, which were not necessarily driven by solid and authoritative land titling schemes, exacerbated conflicts between land claimants: Indigenous communities, RN fighters, and Sandinista officers. In addition, Sandinista and RN soldiers who remained in the Bosawas area at the end of the 1980s civil war and who had received nothing upon demobilization also required access to land and livelihoods. The creation of the Bosawas Biosphere Reserve added the claim of the national government to those of local groups.

The opposing claims that arose out of the creation of a frontier in the Bosawas area led inevitably to conflict. Perhaps the most obvious remedy would have been for the central government to develop an effective and forceful presence to guarantee land titles they granted and to secure order. However, such a presence could only succeed at great cost. Government colonization projects were common in the past, but have become less popular in recent years, mostly because they are expensive and difficult to maintain, but also because of pressure from environmentalists. Moreover, this process would have taken time and a stable, secure government. The conditions during the 1970s and 1980s were such that the Chamorro government in the 1990s did not have the leverage and influence to undertake organized settlement of the area with secure land tenure and titles for claimants.

In fact, however, simply imposing an external, centralized model of government might not have been sufficient. For example, the attempt on the part of the Sandinista government of the 1980s to integrate indigenous communities in the RAAN more closely into Nicaragua resulted in resounding failure, mostly because it failed to consider a history of relative sovereignty in the area in addition to marked cultural differences. As discussed, establishing effective government at the frontier is vital to maintaining peace and sustainability, but this government must absolutely be a combination of "emerging" government, which represents the interests of parties at the frontier, with "imposed" government, which provides the backing, resources, and authority of a central government.

Ultimately, it seems unlikely that the central Nicaraguan government would have been able to govern an orderly resettlement, even assuming it had been willing to make the investment. Given this shortcoming, encouraging resettlement by giving away land starts to look like a bad idea, a quick fix to the problem of reintegrating the combatants of the 1980s. The Chamorro government's answer to creating a satisfactory peace between Contras and Sandinistas was to distribute land in areas where it was apparently available and to preserve the area's biodiversity by creating a park. In fact, this approach was not suitable for promoting peace. By creating land disputes between ex-Contras, ex-Sandinistas, and indigenous groups, this policy led to the formation of the FN 3-80, the FUAC, and the YATAMA within a very short time after the supposed final peace settlement.

Undoubtedly, the negotiated peace settlement between Sandinistas and Contras at the end of the 1980s required provisions for retiring fighters. However, reintegration into Nicaraguan society required more than promises of land. For example, at one point the FUAC negotiated college scholarships for its members; this could have been a productive way of both furthering peace and developing a more educated, qualified professional class. Credit, say, for the establishment of commercial enterprises, might also have been a viable alternative. However, land distribution during the resettlement process and in peace negotiations with splinter groups was, from a frontier point of view, not a viable way to contribute to lasting peace.

Undoubtedly, the national government's claims to land in this area in the form of the Bosawas Biosphere Reserve also contributed to land scarcity and conflict. It would be unproductive, indeed destructive, to say that governments and international organizations should not attempt to start or continue conservation efforts during periods of conflict. However, they must approach these efforts with more careful, considered planning than did the Nicaraguan government in this case. Undoubtedly, the creation of parks can add to the number of participants in existing disputes, potentially exacerbating conflict. However, this does not necessarily mean that parks and lasting peace cannot both succeed. Research has shown that in situations where parks already exist, training local staff to manage the park themselves, without the assistance of international organizations and national elites, and continuing international funding to parks in conflict situations is an effective way to promote the success of park management during conflict situations (Fimbel & Fimbel, 1997; Lanjouw, 2003). Such constructive management of the park discussed in this case study would certainly have furthered an end to conflict as well.

At the beginning of the 1990s, funds for both park formation and management and for the provision of social services and fulfillment of land promises to settlers was in short supply. The Ministry of Environment and Natural Resources raised money from USAID, GTZ, and TNC for management of Bosawas, but funds for social services and general development continued to be scarce. The central government repeatedly failed to fulfill promises of services, credit, and land. Deprived settlers resented the fact that funds were channeled to the park, aggravating their impression that they were being cheated by a central government that reneged on its commitments. The Nicaraguan government did raise limited funds for social services, but resources made available for conservation and social services essentially contradicted each other. Had Nicaraguan government, or contributing nongovernmental organizations, and multilateral and bilateral development assistance agencies integrated services and support for the people in the area with park conservation programs, park conservation might have contributed to development, and local residents might have been persuaded to support the Bosawas Biosphere Reserve.

As a basic start to this effort, the federal government might have expanded the effort to train decommissioned YATAMA fighters for park administration positions, as the YATAMA requested. In general, participation in park management might give local people a vested interest in the preservation of the park's biodiversity, and perhaps increase the perceived relative conservation value of the area to the residents of the region. Additionally, dissemination of power into the hands of local populations would avoid those cultural clashes experienced by the Sandinista government officials when the government attempted to create a presence in the area.

Unfortunately, the government, due to its involvement in the conflict and its own land claims, was neither able nor eager to analyze the long-term effects and effectiveness of either the distribution of land as a way to make peace or the creation of the park. International organizations (NGOs and GOs), as relatively impartial parties, should have stepped in to point out that attempts at conflict resolution were grossly inappropriate and should have provided support and assistance in designing a more appropriate strategy for building peace in the Bosawas region, including emphasizing the importance of establishing a national park well-rooted in the peace effort. Had these organizations attempted to create a positive influence in the area, in the form of an integrated development and conservation program, they could have helped avert some of the conflict of the 1990s. Although maintaining a presence in regions afflicted with violent conflict is a chronic problem for the international community, an early return at the end of the 1980s' civil war with support and funding for the newly elected Nicaraguan government would have helped Nicaragua achieve a lasting peace. The need for increased international participation and support in the Bosawas Biosphere Reserve area continues to this day.

From an analysis of the situation in Nicaragua, one thing is certain: Creating a set of guidelines or rules for dealing with frontier conflict would be impossible because each situation arises from a unique set of circumstances that demand individual attention. Governments and other national and international organizations must take great care in planning approaches to governance and conflict resolution in frontier regions, considering the needs and demands of local groups, and trying to develop a local government in which citizens can govern themselves within the context of existing national government structures.

REFERENCES

Alston, L. J., & Mueller, B. (2003). Property rights, violence, and the state. *Série Textos para Discussão,* No. 293. Brasília: Universidade de Brasília, Departamento de Economia.

Bilsborrow, R. E. (2002). Migration, population change, and the rural environment. *Environmental Change and Security Project Report, 8*, 69–94.

Blom, E., Bergsmans, W., Dankleman, I., Verweij, P., Voeten, M., & Wit, P. (Eds.) (2000). *Nature in War: Biodiversity Conservation During Conflicts*. Mededelingen No. 37. Amsterdam: Netherlands Committee for International Nature Protection.

Browder, J. O., & Geoffrey, B. (1997). *Rainforest Cities: Urbanization, Development and Globalization of the Brazilian Amazon.* New York: Columbia UP.

Carr, D. L. (2004). Proximate population factors and deforestation in tropical agricultural frontiers. *Population and Environment, 25*, 585–612.

Chomitz, K. M., & Gray, D. A. (1996). Roads, land use, and deforestation: A spatial model applied to belize. *World Bank Economic Review, 10*, 487–512.

Collier, P., & Hoeffler, A. (2001). *Greed and Grievance in Civil War*. World Bank Paper No. 31. Washington, DC: World Bank.

Cuadra, E., & A. Saldomando. 1998. *Pacificación, Gobernabilidad y Seguridad Ciudadana: Orden Social y Gobernabilidad en Nicaragua 1990–1996.* Managua: CRIES. 105–138.

Dudley, J. P., Ginsberg, J. R., Plumptre, A. J., Hart, J. A., & Campos, L. C. (2002). Effects of war and civil strife on wildlife and wildlife habitats. *Conservation Biology, 16*, 319–329.

Fearnside, P. M. (2001). Soybean cultivation as a threat to the environment in Brazil. *Environmental Conservation, 28*, 23–38.

Fearnside, P. M. (2002). *Avança Brasil*: Environmental and social consequences of Brazil's planned infrastructure in Amazonia. *Environmental Management, 30*, 735–747.

Fearnside, P. M., Laurance, W. F., Albernaz, A. K. M., Vasconcelos, H. L., & Ferreira, L. V. (2005). A delicate balance in Amazonia: Response. *Science, 307*, 1044–1045.

Fimbel, C., & Fimbel, R. (1997). Rwanda: The role of local participation. *Conservation Biology, 11*, 309–310.

Flores, J. (1998). *Indigena Muere en Refriega: YATAMAs Secuestran a Soldados en Bismuna.* La Tribuna.

Global Witness. (1995). *Forests, Famine and War: The Key to Cambodia's Future*. London: Global Witness.

Global Witness. (1998). *A Rough Trade: The Role of Companies and Governments in the Angolan Conflict*. London: Global Witness.

Global Witness. (1999). *A Crude Awakening: The Role of Oil and Banking Iindustries in Angola's Conflict.* London: Global Witness.

Global Witness. (2002a). *All the President's Men: The Devastating Story of Oil and Banking in Angola's Privatized War.* London: Global Witness.

Global Witness. 2002b. *Logging Off: How the Liberian Timber Industry Fuels Liberia's Humanitarian Disaster and Threatens Sierra Leone.* London: Global Witness.

Godoy, R., Jacobson, M., DeCastro, J., Aliaga, V., Romero, J. & Davis, A. (1998). The role of tenure security and private time preference in neotropical deforestation. *Land Economics, 74*, 162–170.

Homer-Dixon, T. (1999). *Environment, Scarcity and Violence*. Princeton, NJ: Princeton University Press.

Hotte, L. (2001). Conflicts over property rights and natural-resource exploitation at the frontier. *Journal of Development Economics, 66*, 1–21.

Immigration and Refugee Board of Canada (IRBC). (1997). Nicaragua Update. Ottawa: Research Directorate of the Policy, Planning, and Research Branch.

INETER (Instituto Nicaragüense de Estudios Territoriales). (1999). República de Nicaragua: Mapa de la División Politico-administrativa. 1: 750,000. 9 Sept 2005. http://www.ineter.gob.ni/metadatos/imagenes/mapadpolitico.htm.

Kaimowitz, D., & Fauné, A. (2003). Contras and comandantes: Armed movements and forest conservation in nicaragua's bosawas biosphere reserve. In S. V. Price (Ed.) *War and Tropical Forests: Conservation in Areas of Armed Conflict* (pp. 21–47). New York: Haworth Press, Inc.

Koutouzis, M. & Perez, P. (1996). *Atlas Mondial des Drogues*. Paris: PU de France.

Lanjouw, A. (2003). Building partnerships in the face of political and armed crisis. In S. V. Price (Ed.) *War and tropical forests: conservation in areas of armed conflict* (pp. 89–110). New York: Haworth Press, Inc.

Lopez, V. (1999). *Dicen que Ahora si se Desalzaron los YATAMAs*. El Nuevo Diario: 2.

Ministerio del Ambiente y los Recursos Naturales de Nicaragua (MARENA). (2001). Bosawas en Nicaragua. 9 Sept 2005. http://www.tmx.com.ni/bosawas/images/mapa_nicaragua_gif_2.gif.

McNeely, J. (2003). Biodiversity, war and tropical forests. In S. V. Price (Ed.) *War and tropical forests: Conservation in Areas of Armed Conflict* (pp. 1–20). New York: Haworth Press, Inc.

Merrill, T. (1993). *Nicaragua: A Country Study*. Washington, DC: Federal Research Division, Library of Congress.

Nagel, B. Y. (1999). Unleashing the fury: The cultural discourse of rural violence and land rights in paraguay. *Comparative Studies in Society and History, 41*, 148–181.

Nepstad, D., McGrath, D., Alencar, A., Barros, A. C., Carvalho, G., Santili, M., & Vera Diaz, M. del C. (2002). Frontier governance in Amazonia. *Science, 295*, 629–631.

NicaNet. (2002). PLC Sweeps Atlantic Coast Elections but YATAMA-FSLN Alliance. March 11 2002. http://www.nicanet.org/hotline.php?id=16.

NicaNet. (2003). At last! Genuine Autonomy for the Atlantic Coast. July 7 2003. http://www.nicanet.org/hotline.php?id=73.

Rose, C. M. (1994). *Property and Persuasion: Essays on History, Theory, and Rhetoric of Ownership*. Boulder, CO: Westview Press.

Turner, F. J. (1893). The significance of the frontier in American history. In Turner, F. J. (Ed.) *The Frontier in American History,* (pp. 1–27). New York: Henry Holt & Co.

Schneider, R. R. (1995). *Government and the Economy on the Amazon Frontier.* World Bank Environment Paper No. 11. Washington, DC: World Bank.

Whitmore, T. C. (1999). Arguments on the forest frontier. *Biodiversity and Conservation, 8*, 865–868.

World Rainforest Movement (WRM). (2000). Nicaragua: Indigenous peoples' rights and the Bosawas reserve. *World Rainforest Movement Bulletin, 38*, 11–12.

CHAPTER 5

PAST CONFLICTS AND RESOURCE USE IN POSTWAR LAO PDR

YAYOI FUJITA, KHAMLA PHANVILAY, AND DEANNA DONOVAN

1 INTRODUCTION

Forests in mainland Southeast Asia have throughout history been affected by regional conflict, but perhaps more so in the modern era (see Chapters 2 and 9). In the Indochina War, which engulfed this area for several decades during the middle of the twentieth century, forests became variously not only a buffer between hostile groups and a refuge for both civilian populations and combatants, but also a battlefield. The war between Vietnamese and American forces has had both direct and indirect impacts on forests in Lao PDR.[1] Because forests provided for some the cover and resources critical for survival and for others marketable goods, the forest became a target of military attack. Extreme conflict in this region thus resulted in fragmented forests in some areas and impeded forest management in almost all areas, leaving vast tracts of forest as virtually an open access resource. The government's lack of a coherent forest management plan and the difficulty of exercising control over forest resources resulted in widespread, unregulated timber harvesting. Consequently, tackling forest management problems became a primary concern for the postwar government.

A landlocked country in the heart of Southeast Asia, Laos is home to approximately six million inhabitants (2004 est.). Forests, the country's major natural resource, play a significant role in both the national and local economies. In the early 1960s forests covered 17 million ha and 64% of the country. A nationwide survey in the early 1990s, however, revealed that forest cover had fallen to 47% or 11.2 million ha (NOFIP, 1992). The alarming loss of rich natural forest is not only of global concern but also has serious implications for the national economy and livelihoods, especially given that so many rural inhabitants are highly dependent on forest resources for daily subsistence needs as well as cash income.

[1] Lao People's Democratic Republic, hereinafter referred to as Laos.

W. de Jong, et al. (eds.), Extreme Conflict and Tropical Forests, 75–91.

Figure 5.1. *Migration to Vientiane, 1960–1970.*
Source: Based on Whitaker et al. (1971).

The period of the Indochina War,[2] indeed, ever since World War II, has been a time of massive population movement throughout mainland Southeast Asia but especially in Laos (Goudineau, 1997). New settlements have been established and the area of agricultural land expanded as people moved from one region to another for most of the 1960s and 1970s (Goudineau, 1997; Whitaker *et al.*, 1971). Northwest Vientiane Province in particular experienced a large influx of population throughout the period of the war as migrants from the north moved into this area. Forests in the Phou Phanang and Phou Phadaeng mountain ranges located west of National Route 13 in Vientiane Province served both as a military base and buffer between warring forces (Figures 5.1 and 5.2). The influx of war refugees to the Phou Phanang area continued until 1975. Since the installation of the new government in 1975, increased government concern over forest degradation and environmental problems has provoked restrictions on resource and land use in the Phou Phanang area. These problems were

[2]Considering both regional and civil wars, the period of conflict lasted from 1953 to 1975.

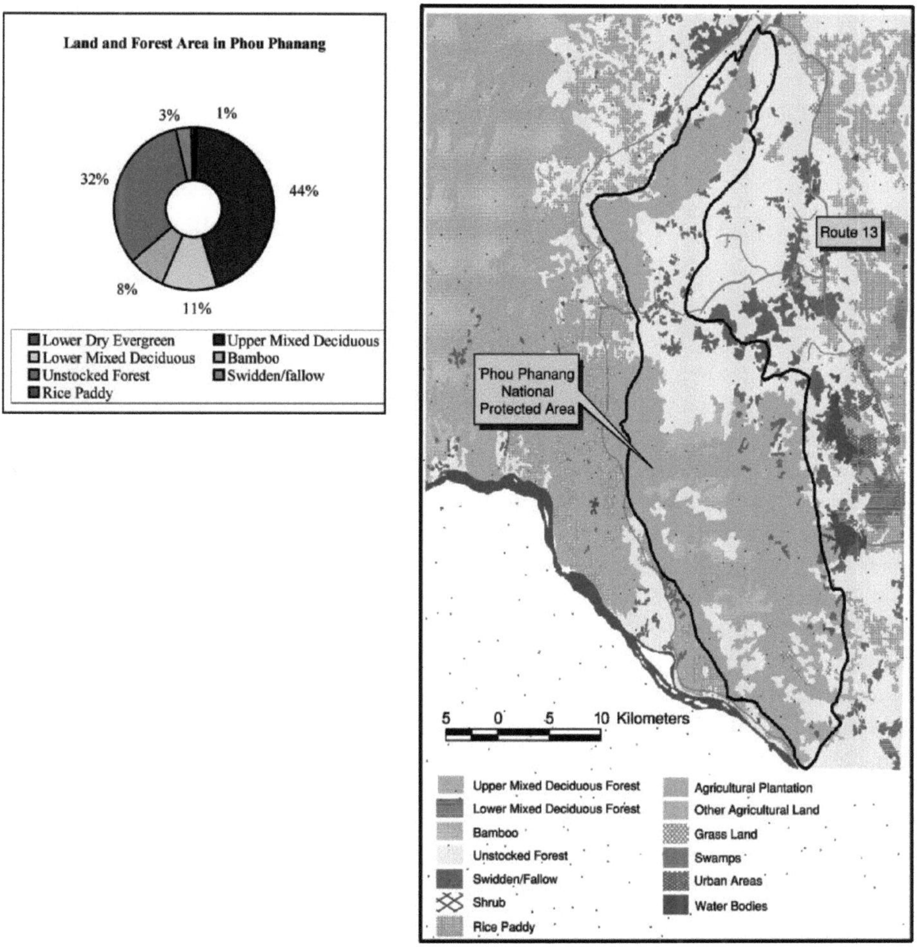

Figure 5.2. *Land and forest cover in Phou Phanang.*
Source: Based on Land and Forest Cover Map (NOFIP, 1992).

exacerbated, however, with the introduction of the protected area system in 1993. The Phou Phanang and Phou Phadaeng mountain ranges became a designated part of the National Protected Forest Area or Phou Phanang Pā Saguan Hēngxāt (hereinafter Phou Phanang or Phou Phanang National Reserve Forest). With this designation, the government defined forest resource boundaries and prohibited human intervention to protect biodiversity.

This chapter focuses on resource-use problems in Phou Phanang that are in large part a legacy of the war. In particular, it focuses on the apparent discrepancies in resource access by the several villages located in areas surrounding Phou Phanang National Reserve Forest. Intervillage differences in resource access observed in Phou Phanang indicate that wartime displacement and postwar resettlement of citizens in this region have had a significant impact on subsequent land-use patterns and livelihood conditions. Understanding current forest and land management problems in Phou Phanang requires a careful consideration of village history and local villages' success in adapting resource use and management practices to changing demographic and politico-economic conditions.

2 FOREST AND MIGRATION DURING THE INDOCHINA WAR

The problem of forest loss and degradation in Laos was first raised as a national concern in 1975, after establishment of the new socialist government. Extensive forests had been destroyed during the Indochina War as forests came under attack for providing not only shelter for refugee populations and a buffer zone between hostile forces, but also cover for combatants. For example, during the 1960s and 1970s forces supported by the United States cleared areas of forest in the present-day Xaysomboun Special Zone (Figure 5.1) to provide a line of defense against the Pathet Lao.[3] Significant forest loss also occurred when the Royal Lao Government constructed the Nam Ngum reservoir north of Vientiane (Figure 5.1). Inundating approximately 37,000 ha of forest land, reservoir construction was strategically located to check the advancement of the Pathet Lao forces and to directly affect areas under their control. Thus, a vast tract of forest was cleared and an enormous number of people were displaced from their villages. A significant proportion of migration occurred between 1960 and 1970 when the United States intensified bombing in northeastern Laos. A record number of refugees, an estimated 730,000, had fled to Vientiane by the end of 1970s. A great majority of the refugees who migrated to the Vientiane region settled in makeshift villages established in or near the forested areas.

During the 1960s forest management received increased attention. The Lao government's American advisers had concluded that the timber industry had significant export potential and began to promote log production (Tsechalicha & Gilmour, 2000; USAID, 1969). The United States, Canada and the Royal Government of Laos jointly carried out a forest inventory in the late 1960s to calculate the commercial potential for timber extraction. During this period and well into the 1970s, commercial logging was widespread. While the Indochina conflict inhibited extensive exploitation of forest resources, unregulated logging and the export of sawn wood was nevertheless rampant, particularly in areas along the Mekong River (USAID, 1969).

2.1 Recent History of Forest Management

Postwar forest management in Laos can be divided into two eras: the first between 1975 and the late 1980s and the second since the late 1980s. In an effort to control forest

[3] The Pathet Lao, a communist-inspired, nationalist political movement, was ultimately victorious in the Laotian civil war, which began in the early 1950s and ended with the abdication of the king and the fall of the coalition government in November 1975.

loss and degradation, the new government from 1975 identified swidden cultivation in the uplands as the major cause of forest degradation (Souvanthong, 1995). In the mid-1980s, the government proposed to transform upland agriculture into more intensive production and called for the resettlement of the upland population. This initiative became the precursor to an agricultural land-use stabilization program (Tsechalicha & Gilmour, 2000). The government also established state-run forest enterprises throughout the country and promoted commercial logging and wood export to increase foreign currency earnings (Persson, 1983). Military forces continued to occupy and control most forest areas throughout Laos, ostensibly for security reasons (Chandrasekharan, 1978). Meanwhile, local authorities had the autonomy to decide the volume of commercial logging to be conducted in the provinces. Timber was not only traded commercially but also shipped to neighboring Vietnam and China in compensation for assistance and loans that had been provided during the Indochina War (*Vientiane Times*, July 30–August 2, 1999).

During the second era of postwar forest management the state began to increase its control over forest resources. In the mid-1980s, the government dismantled state and provincial forestry enterprises charging inefficient management. Logging rights were limited to three state-owned companies organized under the Ministry of Defense. In an attempt to control unregulated logging, the government issued annual logging quotas for each province. The first comprehensive national forest survey, conducted during the late 1980s and through the early 1990s, revealed that the country had lost nearly half of its forests in the preceding five decades—a period dominated by civil and regional conflict. Forest now covered 47% of the nation's total land area compared with more than 80% in the 1930s. As a result of these findings, a new land- and forest-use classification system was proposed as the basis for a new management regime.

The area of forest lost in the north and central regions were nearly equal (about 200,000 ha). Areas identified as "potential forest" (or secondary forest) increased in the central and southern regions by 130,000 ha, signifying a high degree of forest degradation in these regions. The forest survey identified swidden cultivation as the main cause of forest degradation, but also highlighted forest fires, unplanned logging, and expansion of infrastructure and rural development as significant contributors (NOFIP, 1992). In the late 1990s, the Mekong River Commission together with GTZ conducted its own study on forest change and concluded that forest degradation in Laos is caused not by swidden cultivation alone but by a combination of logging and swidden agriculture (Tsechalicha & Gilmour, 2000).

2.2 Politico-Economic Pressures Behind Recent Population Movements

Human migration has occurred throughout history in this montane region of mainland Southeast Asia in which Laos is located. Pressures of tribute, taxation, conscription, and corvée as well as banditry weighed heavily on the population probably well until the middle of the twentieth century (Donovan, 2003). In a study on the distribution of ethnic groups in northern Vientiane Province, Iwata (1959) concluded from the oral histories collected that spontaneous movement of population in northern Laos has occurred gradually over the course of the past 300 years. It was only during the period of French intervention that the colonial administration attempted to temper this flow and to fix villages, introducing a new system of local administration (Robequain, 1955). Despite these changes, spontaneous migration continued as people sought escape from onerous tax burdens and better access to lucrative trade routes.

In modern history, the Indochina War with related hostilities lasting for almost two and a half decades has been one of the most significant factors triggering recent successive population movements in Laos (Goudineau, 1997; Whitaker *et al.*, 1971). The first surge of migration began in 1954, as the North Vietnamese forces defeated the French in Dien Bien Phu. Residents of Houaphan Province in northeastern Laos, who had fought on the side of the French in this conflict, then fled the area. Fearing reprisal attacks by the anti-French North Vietnamese, they became the first to move south (Hasegawa, 1981; Whitaker *et al.*, 1971). By 1962, approximately 125,000 refugees had fled Xamneua, the capital of Houaphan Province, and eastern Xiengkhouang Province, just south of Houaphan. As the war intensified, people from these two areas gradually moved west and south migrating towards northern Vientiane Province (Figure 5.1).

Escalation of the United States offensive along the Vietnamese border in northeastern Laos during the 1960s and 1970s instigated another outpouring of refugees. By 1963, the total number of refugees arriving to Vientiane Province numbered 128,000. The majority of northern migrants to Vientiane during this period were Hmong people from Xiengkhouang Province. An additional 150,000 refugees arrived from the north during the 1970s. It is estimated that by the ceasefire in 1975 the total number of displaced persons in Laos was approximately 730,000 people (Goudineau, 1997). In addition, more than 300,000 people fled across the Mekong River into Thailand during the early 1970s, the closing years of the conflict (Chanda, 1982; Goudineau, 1997).

3 POSTWAR FOREST PLANNING AND POPULATION MOVEMENT IN PHOU PHANANG

In order to understand the present discord in the Phou Phanang area, it is important to understand the relationship of the different groups to the forest and how and why attitudes may have developed and changed over time. The relationship reflects in large part the change in the character of these two entities: the forest and the population. In this section we explore these changes.

3.1 Phou Phanang National Reserve Forest

In 1993, concern for the conservation of Laos' remaining forest materialized in, among other measures, Ministerial Decree No. 164, which set aside the Phou Phanang area as the Phou Phanang National Reserve Forest. The gazetted area stretches across the Vientiane Municipality and northwest Vientiane Province covering a total of 70,000 ha. According to a 1992 forest- and land-cover map provided by the National Office of Forest Inventory and Planning (NOFIP), the Phou Phanang area consists largely of mixed deciduous forest (44%) and unstocked forest areas (34%) mainly in the north and spreading along the east side of the reserve and parallel to National Route 13 (Figure 5.2). Relatively dense forest remains on the western edges of the Phou Phanang mountain range, while forest in eastern Phou Phanang is highly disturbed with numerous patches of bamboo.

The high incidence of forest degradation in the eastern Phou Phanang can be explained by the area's relatively easy accessibility by roads originally constructed for logging purposes. Commercial logging has been rampant in the areas along the Mekong River and the National Route 13 ever since the 1960s. Areas of forest in Phou Phanang were further exploited for commercial purposes after liberation in 1975. The high concentration of population along National Route 13 has also aggravated forest degradation in this area especially due to increased pressure for agricultural land.

Areas northwest of Vientiane experienced dramatic changes in population during the period of the Indochina War. The Phou Phanang and Phou Phadaeng mountain ranges central to the present-day Phou Phanang National Reserve Forest, northwest of Vientiane, served as a natural barrier to and buffer between the warring forces of revolutionary Pathet Lao and the Royal Lao Government (RLG). On the eastern side of the Phou Phanang, along the National Route 13 North, the RLG forces, with the assistance of the United States military, established bases in areas of the present-day Naxaythong and Phonhong Districts. Until the early years of 1960s, the RLG forces controlled areas along the National Route 13 up to Vang Vieng, which lies north of Phou Phanang, while areas west of the National Route 13 in northwest Vientiane were consolidated by the Neo Lao Hak Xāt (NLHX), an underground liberation movement that first arose in southern areas along the Mekong River in the early 1950s. By 1957, NLHX had consolidated their base in areas west of Phou Phanang in present day Sangthong District from whence they conducted their antigovernment operations (Deuve, 1984). By the mid-1960s NLHX forces had gained a foothold in Phonhong District forcing the RLG forces to retreat south. Though now more than 30 years have passed, these activities extending over nearly two decades influenced population movements in this area with impacts still reverberating through the forest today.

3.2 Types of Migrants and their Settlements

Fujita and Phanvilay conducted field surveys of the local population in 2000 to assess the impact of conflict and especially the affect of refugee movements on the area of the Phou Phanang Nature Reserve. The survey revealed that 19 out of the 40 villages surrounding Phou Phanang were established after 1960. The majority of households in these new villages is composed of migrants and displaced families from the northern provinces.[4] Our data suggest four distinct types of resettlement patterns among migrants into Phou Phanang as described below.

[4] Northern migrants include migrants originating from Phonsaly, Houaphan, Xiengkhouang, Luangprabang, and Xayabouri Provinces.

The first type of resettlement consisted of displaced populations from the northern provinces. These people fled their homes in groups from Houaphan and Xiengkhouang Provinces in northeast Laos and traveled south, arriving to areas north of Vientiane (Figure 5.1). This migration occurred between 1960 and 1970 when the United States offensives against the Pathet Lao and the North Vietnamese forces were most intense. Migrants often revealed having settled in several different areas prior to their arrival at Phou Phanang. They were also "strategically" resettled to areas recommended by the NLHX. Thus, their migration and settlement followed a pattern of gradual expansion by the NLHX (Deuve, 1984).[5] Upon resettlement, the migrants, taking advantage of the vast tracts of forest they found, undertook swidden cultivation, especially as they did not possess any significant agricultural inputs other than their own labor. Hasegawa (1981), who served as an agricultural advisor in Laos from August 1974 to May 1977, observed that upland rice cultivation was the most widespread in areas of northern Phou Phanang. Given villagers' limited access to lowland paddy fields and lack of any alternative means of livelihood, these farmers often had no choice but to practice swidden cultivation in Phou Phanang.

The second type of migrant to settle in Phou Phanang was former military personnel. They arrived to the area also during wartime, especially between 1960 and 1970, and married local women. The origins of these men ranged widely from the northern to the southern provinces. They often settled in lowland villages where they now cultivate paddy fields inherited through the wife's family. In a number of cases, these ex-soldiers became active in village administration after 1975. Some even served as village headman in the postwar period because of their relatively higher level of education. Of all the four groups of migrants in Phou Phanang, this group became economically better off as they gained access to local resources through marriage and had the support of well-established social networks.

A third type of migrant to this region consisted of people who had been removed from their original villages as a result of infrastructure development that occurred during the 1980s. Three agricultural reservoirs were constructed in the areas east of Phou Phanang in Naxaythong District as part of government efforts to promote irrigated rice production in the lowlands. Households were involuntarily resettled into new administrative villages without significant compensation for the loss of household possessions or properties. Their livelihoods were jeopardized as agricultural lands were lost without any other alternative means of livelihood being made available to them. While some households were able to purchase land in distant villages, some families had no choice but to push agricultural frontiers further into areas of open forest in Phou Phanang. Others sought wage labor opportunities, such as logging, an activity that subsequently

[5] Two distinct flows of migrants occurred in areas north of Phou Pha Daeng in Hinheup District. Original villagers in the region, predominantly of the Tai Puan ethnic group, abandoned their villages due to war and resettled in central areas of Phonhong District seeking assistance and protection by the Royal Lao Government Forces. Few of these households have returned to their original villages after the war to claim their former agricultural land. In the meantime, refugees from the northern provinces, mostly Khmu ethnic group, were encouraged by the NLHX to resettle in the abandoned villages.

became illegal in Phou Phanang after the area's designation as a National Reserve Forest in 1993.

The fourth type of migrant to settle in Phou Phanang consisted of recent immigrants from northern provinces including Luangprabang and Xayabouri. Often, these individuals and families originally came to stay with relatives who had migrated to the area earlier. They left their original villages in order to find agricultural land and economic opportunities in an effort to improve their livelihood. With arable agricultural land already in short supply around Phou Phanang, these new migrants often ended up in the local agricultural labor pool or seeking nonagricultural wage labor opportunities outside the village.

These four types of settlers arriving over the last four decades exemplify the movement of people in the areas surrounding Phou Phanang since the 1960s. Their arrival increased the total population of the area significantly during the period of the Indochina War, particularly during the 1960s and 1970s. Evans (1995) also notes that population pressure increased in rural areas of Vientiane Province during the mid-1960s as additional migrants arrived from Thai Issan or northeast Thailand seeking fertile agricultural land in the Vientiane plain. Towards the end of the war, during the mid-1970s, once again the movement of people across Phou Phanang increased as more than 300,000 people fled Vientiane across the Mekong River. Not all migrants who arrived to Phou Phanang continued their flight to the other side of the Mekong however. Having firmly decided against returning to their original villages, these migrants were then resettled into local villages or incorporated into new villages in Phou Phanang. Of the four groups of migrants who settled in Phou Phanang, the first and third groups often clustered together to form a village on their own. Meanwhile, the other two types of migrants, consisting mainly of individuals as opposed to family groups, tended to merge into existing villages either in association with in-laws or with previously resettled family members. The first and third groups of migrants were particularly constrained by limited access to as well as the limited availability of arable land, especially since most of the new villages were located in less advantageous locations. With few alternative means of livelihood available, households of these new villages turned to swidden cultivation and collected saleable products from local forests in order to support themselves. Over the last decade government policy on forest and land management in Phou Phanang has come to restrict household access to forests without providing alternative sources of income. This has had significant impact on the more recent settlers, especially households with limited access to arable land.

4 FEATURES INFLUENCING SOCIAL AND ECONOMIC BOUNDARIES

Different ethnic groups often possess different sets of livelihoods skills and environmental knowledge, which enable them to more or less successfully build a new life in the resettlement areas. Resource availability and access also significantly affect their success in employing these diverse skills and thus in rebuilding their lives and livelihoods. Restrictions placed on access to needed resources cannot only be very stressful for the migrants, but sow the seeds of future conflict, between different ethnic groups, between residents and newcomers, and between citizens and their government. In this section we discuss the background and challenges faced by the different migrant groups.

4.1 Ethnic Composition of the Migrant Groups

Ethnic distribution in villages of Phou Phanang varies widely.[6] Settlers in the older villages, established on the lowland plains on both the western and eastern side of Phou Phanang, are predominantly Lao and other members of the Tai-Kadai ethno-linguistic groups. Presently, these groups occupy mainly areas west of Phou Phanang in Sangthong District along the Mekong River and its tributaries. Their villages date back more than 100 years since establishment in the current area. Livelihoods in these older villages are based essentially on lowland paddy cultivation. Cash-crop production thrives in villages along the Mekong River since market liberalization in the mid-1980s. Given the fertility of land along the river and the easy access to markets afforded by the neighboring river and road systems, these riverside villages enjoy a distinct commercial and economic advantage.

Other ethno-linguistic groups including Khmu (Mon-Khmer) and Hmong (Hmong-Mien) groups arrived in the Phou Phanang area during the period of the Indochina War, particularly between 1960 and 1970 (Hasegawa, 1981; Whitaker *et al.*, 1971). Khmu migrants from Xiengkhouang and Houaphan Provinces are concentrated in areas north of Phou Phanang. They gradually migrated south from villages in the northern provinces in the 1960s reaching Vientiane in the 1970s. Upon their arrival in the Vientiane area, Khmu migrants were resettled by the NLHX into the areas abandoned by other villagers in the Hinheup and Phonhong Districts along the base of the Phou Phanang and Phou Phadaeng mountain ranges. Other Khmu migrants reached areas in southern Phou Phanang. These migrants joined others to form a new administrative village after the war in the area southeast of Phou Phanang. Previously, these areas belonged to older villages located along the National Route 13 in eastern Phou Phanang.

Between 1960 and 1970 Hmong migrants from the northern provinces also arrived in the area and moved across Phou Phanang, establishing makeshift settlements in the uplands. According to the lowland villagers in Phou Phanang, these Hmong migrants gradually moved southwards along the National Route 13, occupying areas of open land in Phou Phanang to the west of the national highway. Occasionally, members of the migrant Hmong group would visit lowland villages to trade forest products for rice. With the change of government in 1975, many of these Hmong migrants fled across the Mekong abandoning their new settlements in Phou Phanang.[7] Those who remained were resettled in new administrative villages along the National Route 13.

[6] The following ethnic groups were identified in field surveys: Khmu, Hmong, Yao, Tai Daeng, Tai Neua, Tai Puan, and Lao.

[7] Other young members of lowland villages along the Mekong used the opportunity to escape political and economic turmoil in Laos during this period.

4.2 Village Boundaries and Livelihood Constraints

Until recently any sense of boundary in these regions was ambiguous. Natural forest areas served not only as a physical boundary and social buffer between the older villages, but also as an open access resource in which neighboring village members could freely collect a wide range of forest products. The lowland villagers, however, seldom entered this area during the wartime as they feared possible conflict with the migrant populations. After the war, this land was allocated to new migrants who remained in the region. The previous villagers, however, often returned to exercise their traditional customary rights to collect forest products in these forests despite the area having been assigned to the new villages.

Following the end of hostilities in 1975, the new government of Laos encouraged all wartime migrants in urban areas to return to their original villages. In the aftermath of war this was deemed essential to a successful recovery. The national economy had contracted severely due to the disruption of commerce, agricultural production and the flow of foreign aid during the war years. However, fearful for their security and of the difficult living conditions they might find on their old villages, wartime migrants remained in Phou Phanang and resisted returning to their old villages. They were finally absorbed into the new administrative villages established by the government, but given limited access to arable land. It was believed that the collective approach to agricultural enterprise promoted in the late 1970s would compensate for any shortfall in agricultural production and ensure adequate food supplies for the country. Soon after market liberalization in 1986, however, collective agricultural production was quickly abandoned leaving the migrant villages with the same old problems: a shortage of arable land and agricultural capital.

In Phou Phanang, new migrant villages established after the 1960s have faced agricultural land-use constraints. A high percentage of these migrants are Khmu, a Mon-Khmer ethnic group, who arrived from the northern provinces during the war. Initially, these villages had limited access to arable agricultural land (Figure 5.3). Agricultural inputs also tended to be low, as villagers had little capital and few options to invest in agricultural inputs or technology to improve agricultural production. Villages located in the east of Phou Phanang in Naxaythong and Phonhong Districts have the lowest area of paddy per capita. A high percentage of households (ranging between 20% and 60% of total households) in these villages experience over six months of rice shortages.

In addition to the limited access to land, new migrant villages faced difficulties adapting to the changes in agricultural policy in the 1980s. When the agricultural market was liberalized in the mid-1980s, state cooperatives were dismantled and the government ceased supporting farmers through agricultural collectives established in each village. Instead, government began to promote private agricultural production by offering agricultural credits and other incentives to individual farmers. While this opened opportunities for those villages with arable land, the villages with limited agricultural land benefited little from the new policy. In northern Sangthong District, for example, lowland villages with access to paddy fields were able to improve their

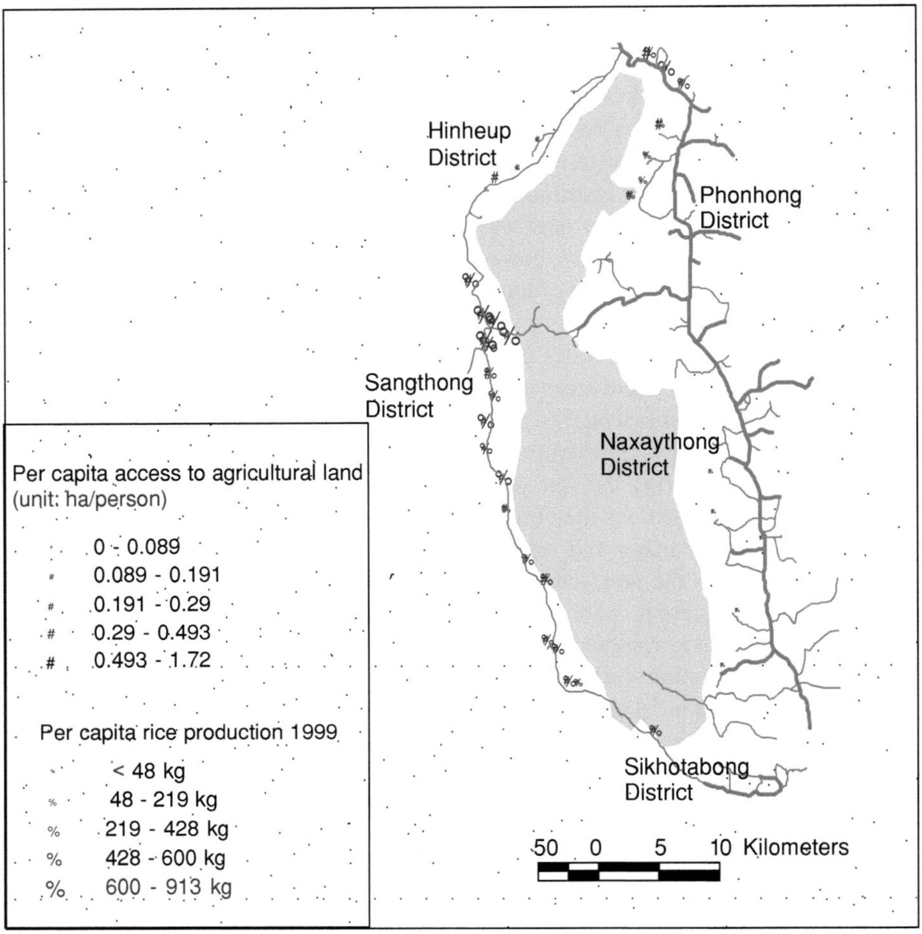

Figure 5.3. *Access to agricultural land and rice production.*
Source: Based on field survey conducted by authors.

rice production through access to formal credit. Meanwhile, other villages in Phou Phanang, particularly the new migrant villages, sought out relatives from whom to borrow money, and in some instances rice, to cover household consumption needs.

Added to these handicaps, new migrant villages were further constrained by the government policy in the 1980s that actively discouraged swidden farming. Development of land and forest management institutions in the 1990s strengthened government control of natural resources as new land-use classifications and new forest boundaries were delineated and mapped. At the same time alternatives to swidden cultivation or other means to improve livelihood conditions in the migrant villages remained limited. Thus, the problems of forest encroachment, poaching and other illegal exploitation of forest resources persisted.

4.3 New Forest Boundaries and Intensifying Conflicts

Following the national forest survey conducted by the NOFIP in early 1990s, a new forest classification system based on forest vegetation type and land cover was enacted, introducing the new notion of resource boundary in Laos. In 1993 the system was also applied to the designation of National Protected Forest Areas as areas of forest were selected and set aside on the basis of their ecological value. The Forest Law promulgated in 1996 also brought a new legal definition to various types of forest cover in Laos, including forest categories such as "production forest," "protection forest," "conservation forest," "regeneration forest," and "degraded" or "barren forest." The new Forest Law also redefined the role of the Ministry of Agriculture and Forestry (MAF) as the executing agency for forest management. Through the Department of Forestry, MAF became the agency responsible for providing technical support as well as forest management services. In 1996 the enforcement of resource boundaries was further strengthened throughout the country with the National Land and Forest Allocation Policy, which supports a process of delineating village boundaries, followed by land-use classification and land-use planning at the village level.

While the new forest management regime in fact encourages local participation in forest management, conflicts over land and forest resources have, nevertheless, begun to intensify over the past decade. Several reasons are identified for this problem.

First, local perceptions of forest resources and good forest resource-use does not necessarily coincide with legally defined forest and its officially sanctioned use. For example, conservation forest, or *"Pa Saguane"*, is a newly defined term under the Forest Law that signifies forest areas reserved for wildlife and wildlife habitat, as well as for endangered tree and plant species. Its protection and management is based on scientific criteria that rules out human use of these resources. Traditionally, villagers set aside areas of the forest to be reserved for future use. In such areas, under customary rules villagers have succeeded in restraining excessive logging, rather than ruling out human intervention altogether. While customary practices as such are not based on written rules, these practices have a certain moral authority and villagers have long been abiding by them as a matter of course.

Second, households may register swidden land through the Land and Forest Allocation Policy. However, the Forest Law and Land Law delimit the area that can be pioneered for swidden. This, in effect, often forces households to continue swiddening with a shortened fallow period of between 2 and 3 years. This practice only accelerates the decline in soil fertility and reduces agricultural productivity while increasing the labor requirement, especially for weeding (Roder, 1997). While the government intention is to encourage intensive use of agricultural land by attempting to curb traditional swidden practices, the outcomes are further impoverishment of upland farm households with limited arable land (Phantanousy, 1994; Pravongviengkham, 2001).

A third and very significant source of problems for current land and forest management policies and programs relates to the security of village and household resource tenure. Despite official government support for customary rights, traditional resource access is often overridden by government fiat, which favors large government projects (Poffenberger, 2000). This problem is compounded when displaced villagers, inadequately compensated for the loss of property, must seek new land and resources for their livelihood in neighboring areas.

5 CONCLUSIONS

Forest degradation in Laos, which intensified during the period of the Indochina War, has lasted nearly 30 years. While the extent and depth of the specific effects of the conflict on forest degradation and forest loss may require further study, clearly war has had negative impacts, both directly and indirectly, on the forest and its inhabitants. Not only did the forest come under direct military attack, it also became shelter and the source of sustenance of combatants and refugees during and after the war. With war preventing the government from developing any consistent or coherent forest management regime throughout the 1960s and 1970s, the forest became an open access resource for warring forces and refugees alike, as well as for enterprising timber merchants seeking to profit from the lapse in government regulation.

At the end of conflict in 1975, the new government focused forest management activities on controlling swidden cultivation and unregulated logging, viewed as the two main causes of forest degradation. In the first decade, the new government attempted to curb swidden cultivation with a stabilization program and the relocation of upland communities as villagers were encouraged to transform swidden cultivation into permanent agriculture. Meanwhile, in order to boost government revenues and economic activity in general, timber extraction was promoted through state and provincial enterprises established throughout the country. A new period of forest management, begun in the late 1980s, was based on a nationwide forest assessment, with the subsequent introduction of a land and forest classification system with formalized resource boundaries for various forest and land types as well as distinct management measures. Despite development of new forest management institutions throughout the last decade, forest degradation continues to be a problem.

Clearly historic patterns of forest and forestland use continue to influence resource use today. At the current annual rate of forest loss of 0.5%, it is estimated that forest in Laos will be reduced to 30% of total area by 2020. In addition to the ecological impact and the effect on agricultural production in many areas of Laos, such change will also undoubtedly intensify competition for and perhaps conflicts over resources in rural areas in the future.

As the history of Phou Phanang Protected Forest Area illustrates, violent conflict significantly affects forest not only directly, through, for example, efforts to clear forest vegetation, but also indirectly, through the disruption of forest policy implementation and management and the displacement of people, who often attempt to rebuild their lives by drawing on forest resources in resettlement areas. Commercial logging accelerated with the advent of war as the Royal Lao Government was unable to control private logging interests, including in the Phou Phanang area and especially during the 1960s (USAID, 1969). After the war, the government resumed authority and placed the area under the control of the military based State Enterprise No. 9, who continued to pursue commercial logging for export, however, in the absence of a scientific forest management plan. In the meantime, migrants from war-struck northern provinces fleeing villages under intense attack by U.S. allied forces between 1960 and 1970 began arriving in Phou Phanang where they established makeshift villages close to the forest. Lacking access to basic agricultural inputs other than household labor, they resorted to swidden cultivation and forest product collection in Phou Phanang. With government control weakened and customary resource management systems disrupted, the Phou Phanang forest became, in effect, an open access resource.

Thus, violent conflict that prevailed during the years of the Indochina war destroyed forests in Laos by several means. After the end of hostilities in 1975 the new Lao government attempted to regain control of forest areas. However, neither did the government policies of the post-1975 era nor those of the post-1985 era alleviate the problem of resource scarcity or the problem of new migrants' limited access to arable land and agricultural capital. Instead, government policies tended to benefit well-established, well-connected lowland villages that increased agricultural productivity through the application of capital-intensive technology and adapted readily to the developing market environment. Migrant villages with limited access to land and agricultural capital inputs and dependent on forest resources to supplement income, on the other hand, have been constrained by the implementation of new government policies on forest and land management which limit access to forest resources.

Differential resource access exacerbated by the incidence of war and the associated population displacement continue to affect households in migrant villages in Laos. Government efforts to discourage swidden cultivation are being strengthened through the introduction of land-use classification and the delineation of resource boundaries. Given that few economic alternatives are open to migrants, the fact remains that Phou Phanang continues to be, in effect, an open access forest and resource conflicts persist. This example serves to illustrate the postwar difficulties of shifting to a new management regime and enforcing new regulations without due consideration of traditional patterns of resource use and the constraints of local villagers. Such problems highlight the need for a careful consideration of the different, unique characteristics of each village, whether of long-term inhabitants or recent arrivals, in particular with regard to their resource access and use, in designing forest and land-use programs and policies. It also illustrates the importance of having a good understanding of local history, and especially recognizing the significance of the impact of violent conflict, both in the short term as well as the longer term, on the forest environment and the local people. By seemingly failing to take into account the impact of recent history, especially the impacts, short- as well as long-term effects of war, recent government actions appear to have intensified the real and potential discord between old and new villages, uplanders and lowlanders, and different ethnic groups over scarce resources and livelihood opportunities.

REFERENCES

Chanda, N. (1982). Economic changes in Laos, 1975–1980. In M. Stuart-Fox (Ed.), *Contemporary Laos* (pp.116–128). New York: St. Martin's Press.

Chandrasekharan, C. (1978). *General Information about Forests and Forestry in Lao People's Democratic Republic.* Bangkok: Food and Agriculture Organization of the United Nations, Regional Office for Asia and the Pacific.

Deuve, J. (1984). *Le Royaume du Laos 1949–1965: Histoire Événementielle de l'Indépendance à la Guerre Américaine.* Paris: École Française d'Extrême-Orient.

Donovan, D. (2003). Trading in the forest: Lessons from Lao history. In Lye, T. P., W. de Jong, & K. Abe (Eds.), *The Political Ecology of Tropical Forests in Southeast Asia—Historical Perspectives* (pp. 72–106). Kyoto: Kyoto University Press.

Evans, G. (1995). *Lao Peasants under Socialism and Post Socialism.* Bangkok: Silkworm Books.

Goudineau, Y. (1997). *Resettlement and Social Characteristics of New Villages: Basic Needs of Resettled Communities in the Lao PDR, an OSTROM Survey.* Vientiane: United Nations Development Programme (UNDP).

Hasegawa, Y. (1981). *Laos Vientiane Heiya, Its Nature, Society and Economy.* Tokyo: Institute of Economic Development of Asia.

Iwata, K. (1959). Ethnic groups in the valley of the Nam Song and the Nam Lik—Their geographical distribution and some aspects of social change. *Minzokugaku Kenkyu, 23* (1–2), 1–29.

National Office of Forest Inventory and Planning (NOFIP). (1992). *Forest Cover and Land Use in Lao P.D.R., Final Report on the Nationwide Reconnaissance Survey.* Vientiane: National Office of Forest Inventory and Planning, Department of Forestry, Ministry of Agriculture and Forestry.

Persson, R. (1983). *Forestry in Laos.* Vientiane: Swedish International Development Agency (SIDA).

Phantanousy, B. (1994). The experience of the shifting cultivation stabilization program of the Department of Forestry. In D. Van Gansberghe & R. Pals (Eds.), *Shifting Cultivation Systems and Rural Development in the Lao PDR.* Vientiane: UNDP/MAF & Nabong Agricultural College.

Poffenberger, M. (Ed.) (2000). *Communities and Forest Management in Southeast Asia.* Gland, IUCN.

Pravongviengkham, P. (2001). Upland natural resources management strategies and policy in the Lao PDR. Paper presented at *International Seminar on Changing People-Environment Interactions in Contemporary Asia: An Area Study Approach,* CSEAS/ASAFAS, Kyoto University, Kyoto, Japan, Nov 15–17, 2001.

Robequain, C. (1955). *L'Evolution Economique de l'Indochine Française.* Tokyo: Yuhikaku.

Roder, W. (1997). Slash and burn systems in transition: Challenges for agricultural development in the hills of northern Laos. *Mountain Research and Development, 17*(1), 1–10.

Souvanthong, P. (1995). Shifting cultivation in Lao PDR: An overview of land use and policy initiatives. Vientiane, Lao PDR: Shifting Cultivation Stabilisation Office, Department of Forestry.

Tsechalicha, X., & Gilmour, D. A. (2000). *Forest Rehabilitation in Lao PDR: Issues and Constraints.* Vientiane, Lao PDR: International Union for the Conservation of Nature and Natural Resources (IUCN).

United States Agency for International Development (USAID). (1969). *Reconnaissance Survey of Lowland Forests: Laos*. Vientiane: Canadian International Development Agency (CIDA) and USAID.

Whitaker, D. P., Barth, H. A., *et al.* (1971). *Laos, a Country Study*. Washington, DC: Foreign Areas Studies, The American University.

CHAPTER 6

SEEING THE FOREST FOR THE TREES: TROPICAL FORESTS, THE STATE AND VIOLENT CONFLICT IN AFRICA

LARRY A. SWATUK

> Forestry issues and opportunities should be examined in a holistic and balanced manner within the overall context of environment and development, taking into consideration the multiple functions and uses of forests, including traditional uses, and the likely economic and social stress when these uses are constrained or restricted, as well as the potential for development that sustainable forest management can offer.
>
> —Preamble (c), Rio *Statement of Principles on Forests*

> In a world of inter-connected threats and opportunities, it is in each country's self-interest that all of these challenges are addressed effectively. Hence, the cause of larger freedom can only be advanced by broad, deep and sustained global cooperation among States. The world needs strong and capable States, effective partnerships with civil society and the private sector, and agile and effective regional and global intergovernmental institutions to mobilize and coordinate collective action.
>
> —UNDP, *In Larger Freedom*, 2005

1 INTRODUCTION

Much of the world's remaining undisturbed moist tropical rainforest is in Africa. It is home to complex, interlinked human and nonhuman forest communities, a great carbon store, and a site of enormous biodiversity. It is also threatened—by population pressure, rapacious logging companies in league with corrupt governments and other groups, and widespread violent conflict. In defense of the forest, various NGOs have demonstrated the relationship between violent conflict and deforestation. For example, Liberia's Charles Taylor used revenue from the sale of timber to directly support the Revolutionary United Front (RUF) rebel movement in Sierra Leone (see Chapters 3 and 7 by de Koning and Price *et al.*). Others have followed the causal pathway in the opposite direction, suggesting that deforestation and other forms of environmental degradation lead to violent conflict (Homer-Dixon, 1999; Homer-Dixon & Blitt, 1998).

This chapter examines the connection between violent conflict and deforestation in the African context, with a particular focus on West and Central Africa. At the same

W. de Jong, et al. (eds.), Extreme Conflict and Tropical Forests, 93–115.

time it examines selected recent initiatives and approaches to managing Africa's tropical forests. The chapter argues that while there are tangible impacts on the forests from widespread violent conflict, neither forest-specific approaches nor narrowly articulated governance interventions will make lasting, positive inroads toward forest conservation and sustainable utilization. This is because, in my view, both violent conflict and deforestation are symptoms of broader socio-political and economic pathologies. Most significant among these is the highly destructive interrelationship between weak African state forms (Herbst, 2000) and a hyperliberal world order (Cox, 1996). Successfully addressing both violent conflict and deforestation in Africa must begin by changing a world order setting conducive to their persistence.

The chapter proceeds in six parts. The second section gives a brief description of violent conflict in West and Central Africa. Section three provides some data on the state of Africa's tropical forests, examines some of the economic benefits derived from their exploitation, and describes some of the many pressures upon them. Section four looks at three primary motivators for global intervention: biodiversity, economic development and 'mystic space." The fifth section briefly reviews some current bilateral and multilateral management strategies, while the sixth and final section concludes the chapter with observations about current approaches and ideas about future policy formation.

2 CONFLICT IN AFRICA

There has been an ebb and flow to major armed conflict in Africa in the post Cold War period, with the number varying from 11 in 1990, to three in 1996, 11 in 1998, nine in 2000, and four at the end of 2003 (SIPRI, 2005). Just prior to the turn of the millennium, large scale wars were being fought in or by Angola, Congo-Brazzaville (Republic of Congo), Congo-Kinshasa (Democratic Republic of Congo, DRC), Ethiopia, Eritrea, Rwanda, Somalia and Sudan. Several more "low-intensity" conflicts were underway in Burundi, Chad, Djibouti, Senegal and Uganda (SIPRI, 2000). In early 2006, few if any of these conflicts have been brought to an absolute end while new conflicts are cropping up (e.g., Côte d'Ivoire). Where there is peace, it is tentative at best—despite the continuing presence of UN and African Union peacekeepers, the holding of elections, the implementation of peace accords, and the imposition of Security Council sanctions against belligerents. Most postconflict countries return to civil or other type of war within 5 years of the signing of ceasefire agreements.

The number of deaths directly related to these conflicts is estimated by UNICEF to total somewhere between seven to eight million, of which two million are said to be children. Two million people are thought to have died in Sudan's 20 year civil war, with approximately 180,000 dead and 2 million displaced since 2003 in the Darfur region; 500,000 died during Angola's more than 25-year struggle, with millions today subject to "excess mortality"—i.e., war-induced vulnerability to death by malaria, malnutrition, diarrhea and respiratory infection. Eight hundred thousand died in the 1994 Rwanda genocide, and more than 300,000 since 1993 in Burundi's civil war. Conflict in the eastern DRC is the direct result of these events. At the end of 2004, 2.74 million of the world's 9.2 million refugees were in Africa. Approximately 3.4 million were internally displaced at the end of 2004. Not only are these flows symbolic of regional instability; they also highlight the great pressure displaced populations place on local resources and already poor host countries—Tanzania, for example, is host to 650,000 refugees and Zambia hosts approximately 225,000.

2.1 Central Africa

Many of the states in Africa at war have been pitted against each other in what is often described as Africa's First World War, whose geographical center is the Great Lakes Region of Central Africa. During the 1998–2003 five year period two different alliances formed in the struggle for sovereign power. The Government of the DRC, itself having come to power through violent struggle in May 1997, supported by Angola, Namibia and Zimbabwe, was challenged by primarily two rebel factions, the Rassemblement Congolais pour la Democratie (RCD or Congolese Rally for Democracy) and the Mouvement de Liberation Congolese (MLC or Congolese Liberation Movement), the latter a breakaway group from the RCD. Uganda and Rwanda were also active destabilizing forces in eastern DRC. The RCD, initially supported by the government of Rwanda, remains active in the "lawless east." A peace accord—the Lusaka Accord—was reached in 2003 but implementation remains problematic, partly due to ongoing rebel activity. While Botswana, Zambia and South Africa helped broker the Lusaka Accord, a joint secretariat of the UN and African Union (AU) is currently attempting to initiate an International Conference on Peace, Security, Democracy and Development in the Great Lakes Region. Protocols and projects are being developed around four broad areas of activity: peace and security; democracy and good governance; economic development; and social humanitarian issues. An interministerial conference involving Angola, Burundi, Central African Republic, DRC, Kenya, Republic of Congo, Rwanda, Sudan, Tanzania, Uganda and Zambia is slated for some time following the June 2006 DRC national elections.

Four million people are estimated to have died in the DRC conflict since 1997, leading the UNHCR to describe the situation as the world's worst humanitarian disaster in the last 60 years. Thirty eight thousand people per month are said to perish from excess mortality. Aside from formal refugees, another 800,000 are identified by UNHCR as vulnerable groups, of which 40% are children, and 49% are women. Despite the Lusaka Accord, and the presence of nearly 17,000 UN peacekeepers (through MONUC, extended to end-September 2006), vulnerable individuals "continue to perish silently in the forests."

According to the International Conference on the Great Lakes Region background paper, "The Great Lakes Region has confronted Africa with a new conflict system: interstate conflicts have become the continuation or manifestation of intrastate conflicts (civil wars) by other means, namely, direct military attacks, instigation of and/or support to rebel movements in the target country, which may similarly be engaged in the exercise of destabilization by proxy. These conflicts therefore constitute a form of transnational warfare involving states as well as armed groups with cross-border ties to states, social movements, markets, criminal cartels and even corporations." Western states, individuals and transnational corporations (TNCs) engage in formal (e.g., offshore oil, mining) or informal (e.g., blood diamonds, coltan, conflict timber) exploitation of resources in the region. They pay rents to belligerents who in turn use this income to purchase weapons. Former East European and Soviet Union countries such as Bulgaria, Romania, Russia, and Ukraine supply weapons either directly or indirectly through transshipment from an intermediary party. Intermediaries like Burkina Faso facilitate transshipment of weapons to Central and West African conflicts by appearing as the recipient in end user certificates though they themselves do not participate directly in the fighting. Finally, other African states like Libya and Sudan indirectly participate by training and arming rebels or by providing weapons or cash to rebel groups. One must also include the numerous gray actors in these conflicts—from mercenary groups to aid agencies, from international nongovernmental organizations to local individual fortune seekers. The array of parties actively prosecuting or benefiting from conflict and instability in DRC is truly staggering.

2.2 West Africa

In West Africa, Liberia, Guinea and Côte d'Ivoire are linked geographically through the densely forested Nimba Mountains. For Paul Richards (2005), this bio-hotspot is the location for four of Africa's current or recent civil wars. Liberia's 15-year long civil war (1989–2003) spilled over to its western neighbor, Sierra Leone, in the early 1990s, and to Guinea in early 2001. These conflicts were brought to the world's attention largely through the Western media, which chose to emphasize the particularly barbaric and bizarre nature of warfare there (Ellis, 1999). While peace processes in Sierra Leone and Liberia seem to have begun to bear fruit through relatively open elections in 2003 and 2005, Côte d'Ivoire has been politically divided since 2002, with rebel groups—Forces Nouvelles—holding the north, and government forces holding the south (see Chapter 3). A 7,000 strong UN peacekeeping mission, UNOCI, has been in place since February 2004. While Ivorian President Gbagbo has been working toward national reconciliation and elections are slated for October 2006, there is little indication that north–south differences are near anything other than temporary resolution.

While the number of deaths directly attributed to the Liberian and Sierra Leonean civil wars possibly totaled less than 90,000 people (Ellis, 1999, pp. 314–316), millions suffer through inter alia infrastructure destruction (leading to loss of economic opportunity, rising levels of poverty, and vulnerability to otherwise preventable diseases), internal displacement, deliberate physical maiming by rebel groups, and HIV/AIDS. As with Tanzania and Zambia, weak West African states continue to bear the heavy burden of supporting hundreds of thousands of refugees. Although former warlord and Liberian President Charles Taylor's agreement to step aside and enter voluntary exile in Nigeria made it possible to begin postwar reconstruction, many feel that Taylor still wields great influence over the course of events in West Africa. Now that

Ellen Johnson–Sirleaf has taken power in Monrovia and Taylor has been arrested, questions are being raised regarding how to bring Taylor—charged with 17 counts of war crimes and crimes against humanity—to justice without destabilizing Liberia's fragile peace. The fact that the outgoing National Transition Government of Liberia looted their offices upon leaving does not give one much hope for lasting stability in this devastated country (see www.irinnews.org/report.asp).

As with the DRC, the various conflicts that have raged over West Africa in the post-Cold War period have brought together a wide variety of very strange bedfellows: external patrons such as Libya's Gaddafi, Burkina Faso's Compaore, and Côte d'Ivoire's Boigny, before his recent death; several mercenary groups brought in to defend both companies (Executive Outcomes) and governments (Gurkha Security Guards, Executive Outcomes) and even to help with ECOMOG—the regional peacemaking body—logistics (Pacific Architects and Engineers); regional and international banks which laundered millions of dollars; TNCs involved in mining, logging, and plantation agriculture; diamond smugglers; market traders from all over the region; numerous warlords less intent on taking state power than with carving out a piece of the state for themselves; entrepreneurs within ECOMOG itself; and several regional state-makers (Babangida; Rawlings) who feared the domino effect of a Taylor-led victory in the region.

3 SOME DATA ON AFRICAN FORESTS

As this is a comparative examination of tropical forests and violent conflict, it is important at this point to change direction somewhat by constructing a landscape for these conflicts. Tropical forests have played a central role in each of the conflicts—as bases for rebel movements, as contested terrain, as sources of valuable natural resources (rubber, timber, diamonds, gold and other minerals) whose exploitation helped finance state and antistate groups alike—and these forests have, in turn, been fundamentally impacted by the prosecution of conflict. This section, therefore, presents data concerning the extent of tropical forests in West and Central Africa, the contribution of forestry to the economies—and by extension, to the conflicts—of the region, and a brief discussion of the impact of population growth, fuelwood consumption, urbanization and the expansion of agriculture on forest resources.

For some, Africa has become the last forest—environmental—global stand. According to the World Resources Institute (WRI, quoted in UNEP, 1999, p. 38), "eighty per cent of the forests that originally covered the Earth have been cleared, fragmented or otherwise degraded." What remains is "located in just a few places, mostly in the Amazon Basin, Canada, Central Africa, Southeast Asia and the Russian Federation" (UNEP, 1999, p. 38). According to UNEP (1999, p. 57),

> "The forests of Africa cover 520 million ha and constitute more than 17 per cent of the world's forests. They are largely concentrated in the tropical zones of Western and Central, Eastern and Southern Africa. With more than 109 million ha of forests, the Democratic

Table 6.1. *Land and forest area for selected Sub-Saharan African countries.*

Selected countries	*Total land area (000 ha)*	*Forest as a percentage of total area*		*Frontier forest (%)*	*Threat to frontier forest (%)*
		Original forest (%)*	*Current forest** (%)*		
Angola	124,670	19.8	15.3	—	—
Dem. Rep. Congo	226,705	82.5	60.4	15.6	70.4
Congo	34,150	100	67.8	28.7	64.6
Rwanda	2,467	36.1	16.1	—	—
Burundi	2,568	46.3	3.5	—	—
Uganda	19,965	70.0	4.4	—	—
Sudan	237,600	1.2	0.0	—	—
Liberia	9,632	99.6	44.2	—	—
Sierra Leone	7,162	100	9.7	—	—
Cameroon	46,540	80.4	42.4	7.9	97.4
Central African Republic	62,298	56.8	15.9	4.4	100
Equatorial Guinea	2805	95.6	38.4	—	—
Ghana	22,754	65.9	8.6	—	—
Gabon	25,767	100	90.4	32.4	100

Source: WRI, *World Resources* 2000.
*According to WRI, "original forest" is defined as that thought to exist 8,000 years ago. The quantification of this is problematic, but is used for comparative purposes.
**As a percentage of original forest.

> Republic of the Congo alone has more than 20 per cent of the region's forest cover,[1] while Northern Africa has little more than 9 per cent .... Forests include dry tropical forests in the Sahel, Eastern and Southern Africa, humid tropical forests in Western and Central Africa, diverse sub-tropical forest and woodland formations in Northern Africa and the southern tip of the continent, as well as mangroves in the coastal zones."

The WRI (2000, p. 255) identifies 31 Sub-Saharan African countries as having more than 1 million ha of tropical forest. Of these, tropical forest covers more than 50% of the DRC's surface area. The next largest forests are to be found in Angola, the Republic of Congo, Zambia, Gabon, Mozambique and Cameroon (see Table 6.1). All of these countries, save for Zambia (a former British colony), have a history of political instability, with Angola (until 2002) and the DRC involved in constant warfare since they came to independence in 1975 and 1960, respectively.

As can be seen in Table 6.1, forests throughout West Africa, save for Gabon, are badly degraded. Indeed, according to WRI (2000), whereas Sierra Leone was at one time thought to be completely forested, less than 10% of that original forest cover remains. Similarly, only 8.6% of Ghana's original forest remains. Liberia, however, has fared somewhat better, with roughly 44.2% of its original forest remaining as current forest. Among parties to the Central African conflict, areas which were fairly extensively

[1]These numbers vary by as much as 20% depending on the source consulted.

Table 6.2. *Tropical forest area, protected area, and trade in forest products.*

Selected countries	*Total area of tropical forest (000 ha)*	*Proportion of tropical forest protected (%)*	*Value of imports 1996–1998 (106 USD)*	*Value of forest product exports 1996–1998 (106 USD)*	*Export value as proportion of total exports 1997*
Angola	37,564	2.6	4.6	1.2	0.02
Dem. Rep. Congoo	135,071	6.6	6.0	63.3	4.38
Congo	24,321	4.4	1.4	147.4	8.19
Rwanda	291	77	1.9	0.1	0.06
Burundi	219	18.2	2.9	0.1	0.06
Uganda	3,772	17	3.3	0.0	0.0
Sudan	12,288	12.3	34.4	0.1	0.02
Liberia	3,149	2.9	1.0	11.9	n.a.
Sierra Leone	260	20.3	2.2	0.6	0.64
Cameroon	20,009	6.0	22.5	426.6	17.46
Central African Rep.	17,101	20.1	0.2	27.2	12.75
Equitorial Guinea	1,749	0.0	1.0	56.8	13.18
Ghana	1,694	7.1	15.8	144.7	8.73
Gabon	21,481	3.6	4.3	287.6	8.73

Source: WRI (2000).

forested in the past (Angola, Rwanda, Burundi, Uganda) are badly degraded today. Only the DRC and Congo remain extensively forested (Tables 6.1 and 6.2).[2] There is very little frontier forest remaining on the continent, all of it clustered in the Central/West Africa region and all of it deemed seriously threatened (WRI, 2000). Clearly, the fact that this frontier forest exists within and between highly unstable political entities has concentrated the attention of many global actors (Shambaugh *et al.*, 2001).

3.1 War, Shadow Economies and Deforestation

As shown in Table 6.2, international trade in forest products is important for several Central and West African states, either in absolute dollar terms (Cameroon, Gabon, Congo, Ghana) or as a percentage of total exports (Cameroon, Equatorial Guinea, Central African Republic, Ghana, Gabon, Congo). It must be remembered that these are *recorded* exchanges. What is not captured in these figures is the production of the many illegal logging operations underway throughout the continent, many operating along semiclandestine lines similar to Charles Taylor when he was the warlord of "Greater Liberia," and that continued for 6 years during his presidency of the *de jure* Liberian

[2]Whereas the WRI (2000) suggests that roughly 60.4% of original forest remained in 1996, a WRI-commissioned study in the Frontier Forests Initiative suggested that in 1997 86% of original forest cover remained (Wolfire *et al.*, 1998, p. 6). This wide discrepancy is common in attempts both at quantifying existing forests and establishing a benchmark from which to judge current forests with "original forest cover" (see, for example, Fairhead & Leach, 2000).

state. The following quotation is important for two reasons. One, as an example of the kinds of shadow economies existent throughout West and Central Africa today. Two, as a benchmark indicator for environmentalists everywhere who fear that management on a sustainable basis may indeed be beyond their means.

> "The Ivorian port of San Pedro boomed on exports from eastern Liberia, and its main hotel became the base for Liberian exiles, including two ex-ministers who managed their interests in the wood and rubber businesses from this Ivorian base. In 1991-2, Taylor's Greater Liberia is estimated to have produced 343 thousand cubic meters of timber which it exported to France, Germany, the United Kingdom, Italy, the Netherlands, Spain, Greece, Portugal and Turkey. Liberia became France's third-biggest supplier of wood. Among the logging businesses operating in Liberia was a front company which, it is alleged, had close links to the Unification Church, the Moonies. Details of hardwood cargoes carried by just two ships used in the hardwood trade reveal eight shipments of logs exported from Sinoe in southeastern Liberia between December 1994 and March 1995 alone, of which four were to Bordeaux and one to Nantes, in France, and three to Leixoes, Portugal (Ellis, 1999, pp. 165–166)."

Despite the election of Africa's first female President in Liberia in November 2005, little has changed on the ground. Thousands of ex-combatants continue to mine diamonds, engage in pit-sawing activities, and move these resources across borders with impunity—despite UNMIL's mandate "to assist the transitional government in restoring proper management of natural resources" (*An Architecture of Instability*, December 2005, see www.globalwitness.org/reports/).

To be sure, the different order of things facilitated by widespread, violent conflict allows for activities at once unsustainable, unrecordable and highly profitable for those not put off by the risk. Granted, the present context tends to encourage short-term profit maximization or what might be termed fly by night or resource raiding operations. Some of these activities may in fact be sustainable, however. Duffield (2001), among others, suggests that the new medievalism of African politics encourages both long-term, low-level conflict (stalemate) and a sort of quasi-sovereign stability in the warlord zones of influence. Many of these warlords—be they holders or contesters of state-power—have created stable relationships with foreign companies in shared pursuit of their different aims.

For example, Global Witness reported in 2001 that the CEO of Danish forestry company, DLH group, was both a board member of the Liberian Forestry Development Authority (LFDA) and Director of the Royal Timber Company. During his presidency, the LFDA operated as Charles Taylor's private company and was reported to provide him with millions of dollars in nonrecorded extrabudgetary income (see *Taylor Made*, September 2001, www.globalwitness.org/reports/). Taylor used profits from the forestry industry—in the order of USD 100 million/year—to buy arms for the RUF rebels in Sierra Leone. There are indications that he and other rebel groups are still active in Liberia and may be assisting Côte d'Ivoire's Nouvelles Forces to similar ends. The Global Witness report identified France and China as major markets for round wood exports from Liberia, while the UK, Italy, Denmark and Germany purchase significant quantities. Global Witness efforts were influential in leading to both UN Security Council sanctions on Liberian diamond exports (2001) and log exports (2004). However, a recent report (December 2005) suggests that little has changed:

> "Despite the large number of UN forces in the region ... West Africa remains highly unstable. Stability in Liberia itself is only guaranteed through the presence of UNMIL, and the significant problems and failures of the ongoing DDRR [demobilize, disarm, rehabilitate, reintegrate] process in Liberia have created a large population of disgruntled ex-combatants with the capacity and will to remain mercenaries for hire (*Timber Taylor Soldier Spy*, June 2005, www.globalwitness.org/reports/)."

While Blundell (2005) argues that under UNMIL, most logging companies respect sanctions, it is clear that a weak Liberian state in combination with current instability in Côte d'Ivoire provides the perfect setting for continued illegal export of both logs and diamonds in support of various rebel movements and antistate aims.

As with Liberia, so with the DRC. There have been a large number of studies conducted into various types of resource plunder in the Congo. Some of these activities are state-sanctioned, such as the deal brokered between DRC President Laurent Kabila and Zimbabwe's President Robert Mugabe who secured four logging concessions (in Katanga, Kasai, Bandundu and Bas-Coifo Provinces) totaling 33 million ha, or roughly 15% of the DRC, from which he hopes to eventually extract 150,000 m^3 of wood per year. Rather than a personalized agreement, this is reported to be a formal agreement between Comiex-Congo of the DRC and Osleg, a Zimbabwean registered company whose name is an acronym standing for "operation sovereign legitimacy." However, the main actors are high ranking politicians and military personnel, so pointing to personalization of state power throughout much of conflict-torn Africa (see *Mugabe's Loggers to Ravage Rainforest*, August 2001, www.globalwitness.org/reports/). In Eastern DRC, there are also many informal agreements between local authorities, rebel groups, and foreign militaries (Rwanda, Uganda) involving resource plunder, be it the open-pit mining of coltan (necessary for pinhead capacitors in mobile telephones), blood diamonds, or clear-cut logging. Observers have described this free for all as a resource curse (Bannon & Collier, 2003) whereby natural resource abundance facilitates the initiation and prolongation of conflict.

3.2 Fuel for the Fire

The forests of Central Africa cover approximately 241 million ha, accounting for 37% of Africa's total forest cover. One hundred thirty million hectare of total forest is regarded as potentially productive, while the rest is not accessible. During 1990–2000, annual forest loss was estimated to be 934,000 ha, 18% of total deforestation in Africa. Forest loss in Cameroon and the DRC accounted for 80% of this total (FAO, 2003). Despite these totals, deforestation in the conventional sense is not a major issue in Central Africa (FAO, 2003). Nevertheless, Wolfire *et al.* (1998) claim that "outside the seasonally flooded central basin, where most intact forest is located, there is increasing pressure on the resource."

Part of the pressure on land and forest resources derives from trends in population settlement and movement. In terms of local markets, both timber and nontimber forest products are extremely important throughout Africa. According to UNEP (1999, p. 38), "In Africa, where 90 percent of the population depends on firewood and other biomass for energy, the production and consumption of firewood and charcoal doubled between 1970 and 1994 and is expected to rise by another 5 per cent by 2010 .... The total volume of wood exploited annually in the subregion is more than 200 million cubic meters." "Nearly 90 per cent is consumed as firewood and charcoal, and only 2 per cent as industrial round wood" (UNEP 1999, pp. 57–58). In the words of Raskin and Lazarus (1989), the energy crisis in Africa, is a wood fuel crisis. Throughout Sub-Saharan Africa, the continent has experienced not a shift from traditional to modern fuel sources, but backward fuel switching. As a result, demand for fuelwood is increasing in urban, peri-urban and rural areas (FAO, 2003).

To some, "fuelwood scarcity is more a consequence of deforestation than a cause" (Johnson & Chenje, 1994, p. 140). Agricultural expansion over the 1990s led to forest-to-farmland conversion in the order of 200,000 ha per year (UNEP, 1999, p. 58). It is too simplistic to say that this land conversion is a direct consequence of increasing populations needing both more food and more space to grow that food. Rather, it is more accurate to say that deforestation for agricultural expansion is a consequence of the aforementioned, plus extant and increasing inequitable land distribution. It is also a result of a need on the part of states to generate foreign exchange for debt repayment, particularly in the wake of continent-wide structural adjustment programs. Under these programs, more land has come under private leasehold arrangements, for example, large farm and plantation agriculture in Ghana and Cameroon in West Africa and Zambia and Mozambique in Central/Southern Africa. When applied in situations where existing inequalities are quite dramatic—e.g., in countries with large settler communities as in Zambia and Zimbabwe, or where the state is itself a major landholder and cash-crop producer, as in Cameroon and Malawi—land hunger is acute (see Tevera & Moyo, 2000 for an important discussion of the situation in Southern Africa). So, forestland comes under pressure both from out migration and from the more intensive use of lands bordering forested areas. Put simply, there is nowhere else for the poor and disempowered to go. Even forests that used to be managed well by local communities are now suffering under the combined pressures cited above. However, over most of West/Central Africa "the extent of sustainably managed forests is negligible" (FAO, 2003, p. 12).

3.3 *Refuge and Resource*

Lastly, the movement of refugees throughout Central and West Africa, particularly in the Great Lakes region following the Rwandan genocide and the continuing strife in Burundi, has led to widespread degradation of forest resources, particularly at the margins of forests and on the borders of states. Northern Malawi, northeastern Tanzania and eastern DRC have seen dramatic changes to local ecosystems as a result of the influx of sometimes more than a million refugees. Whereas degradation of forest resources in Angola is really quite marginal, the movement of refugees in and out of northern Angola has placed considerable stress on the local environment as refugees and internally displaced people try to find space away from the belligerents. This has not only led to forest degradation but to the so-called empty forest phenomenon whereby people hunt most large mammals to extinction (e.g., for bushmeat, ivory, rhino horn). Internally displaced people and refugee demand for resources aside, the overall market for bushmeat in Central Africa is approximately 1 million tones per annum—equal to four million cattle (FAO, 2003).

4 MOTIVATORS FOR INTERVENTION

As stated at the outset, there are numerous motivators driving attempts at sustainably managing the world's forests. Here I will simply highlight three: biodiversity; economic development; and what I call mystic space.

4.1 Biodiversity

At the global level, two primary concerns with rainforest preservation are climate change and biodiversity. Most often these issues are treated within the contexts of common property resource management or global public goods, legitimating transnational interventions in the management of what are conventionally considered domestic resources. There are several problems commonly associated with global level motivators, perhaps the most salient of which is that attempts at intervention often lead to claims of Western environmental imperialism. Yet, there is no need to resort to global ethical arguments regarding preservation of biodiversity. It is increasingly clear that reduced biodiversity, especially the loss of key species, threatens the sustainability of local ecosystems. The more intact an ecosystem, the better is its resilience in the face of complex and persistent pressure. Part of the argument in support of frontier forests revolves around the importance of unbroken canopy to the stability of the entire ecosystem and its subzones.

> "Central Africa harbors unique endemism, in such areas as the interglacial refuges of the central Congo and Gabon basins, the mangroves of the Atlantic coast, the dry-forest ecozone of the Congo and Zambezi basins, and the lakes and tributaries of the Congo River" (FAO, 2003, p. 11).

The United States government, through CARPE (see below), is protecting a suite of 11 large tracts (or landscapes) of relatively intact wilderness and other areas of unique ecological importance that stretch across the entire length of the Congo Basin, from Uganda in the East to Equatorial Guinea in the West (USAID, 2005). The network of actors interested in preserving Central African rainforest is complex, ranging from a number of civil society and social movement groups (Greenpeace, WWF, Rainforest Action Network, National Geographic Society, CIFOR) to donors (e.g., the United States through USAID's Central African Regional Program for the Environment—CARPE), recipient country organizations (e.g., Congolese Institute for Nature Conservation),

IFIs (World Bank) and IGOs (UNESCO, FAO) and private sector groups (e.g., forestry companies participating in various forest certification programs; Home Depot, IKEA). Many of these are tied together through such programs as the Congo Basin Forest Partnership, the Liberian Forestry Initiative, and the Sustainable Forest Products Global Alliance. Through the Yaoundé Declaration of 1999, and the 2002 Inter-state Forestry Ministers' Working Group, Central African states have declared their intention to work collectively to sustainably manage this shared resource.

There is, however, a seeming lack of urgency. In the DRC, "most of the country's forests are currently inaccessible and of low commercial value, so the opportunity costs in conservation and taking time to plan is low" (Wolfire *et al.*, 1998, p. 9). However, according to Wolfire *et al.* (1998, p. 10), both internal and external pressures "could build rapidly." For example, while logging itself may be sustainable, access roads provide new pathways for the exploitation of nontimber species, the absence of which may ultimately undermine ecosystem stability.

The combination of extensive but weak protected area systems,[3] increased commercial logging and expanding infrastructure could prove lethal. In 1995, for example, the DRC produced 300,000 m^3 of logs of which one-third were exported and two-thirds consumed domestically. When compared with production rates in countries with much smaller forest areas but more stable polities, the untapped potential of DRC becomes clear. This explains the demonstrated interest of transnational logging companies, among them the Zimbabwe Defense Force.[4] While benign selective logging is possible, fear of widespread exploitation is not unfounded.

4.2 Economic Development

Clearly, forestry is seen to be a viable engine of economic development. In Canada, for example, the forestry industry is said to account, directly and indirectly, for one of every seventeen Canadians in formal employment. In Central Africa, there is massive scope for increased production of both timber and nontimber forest products in formal and informal, and modern and traditional sectors (FAO, 2003).

The desire on the part of central governments to enhance control over the resource base and exploit it more fully follows an obvious logic. Central/West African states are among the poorest in the world. In terms of the UNDP's Human Development Index rankings, of 174 countries, the highest ranking is Gabon at 123 with an HDI of 0.592. This last figure is only minimally above the low human development threshold of 0.500. In terms of economic growth, all of these states' formal economies continue to perform poorly. Only Equatorial Guinea has seen a high level of GNP per capita growth in the 1990s, but this is due simply to the windfall gains accruing from the recent discovery and exploitation of oil (UNDP, 2000).

[3] In 1997, protected areas in Central Africa accounted for 36.6 million ha or 6.9% of total land area.

[4] Placing the DRC 21st in world rankings. In comparison, Cameroon produced 2.7 million cubic meters of logs and Gabon 2.1 million cubic meters, placing them sixth and eighth, respectively.

However, all of these states lack the capacity to either exploit or manage these resources effectively, let alone sustainably. Instead, they auction concessions to foreign companies from whom they derive rents. In most cases, these foreign companies are mining the forest, rather than managing it sustainably (see www.globalforestwatch.org). For example, in 1998 there were 479 logging companies registered in Cameroon. This compares with 177 in 1990 and 106 in 1980—the start of the so-called lost decade in Africa. In 1999, 84 individuals and companies held valid documented logging rights. These companies are overwhelmingly European in origin.[5] According to data provided by TREES, in the early-1990s 17.9 million ha of closed forest remained. According to the FAO, Cameroon has lost over half of its historic closed forest cover, including more than two million hectares between 1980 and 1995. Four of the 19 West and Central African deforestation hotspots as identified by TREES are in Cameroon (see www.fao.org).

Global actors recognize the desire on the part of African state-makers to exploit their forest resources, be it for personal or wider social gain. At the same time, African leaders recognize the desire of global actors to exploit the continent's forest resources. A significant challenge for those interested in conserving these resources, therefore, is to determine ways in which such actors generally uninterested in conservation or sustainable utilization can be persuaded to move toward such a position. The means by which this may be accomplished are thought to be increasingly clear—e.g., as laid out in terms of (African) Forest Law Enforcement and Governance (and Trade)—and concerted efforts have been underway since the 2002 Johannesburg World Summit on Sustainable Development (WSSD). However, there has been as much backsliding as there has been progress. African states uniformly face declining capacity in the public sector, generally, and in national forest agencies in particular. At the same time, TNCs are interested in the extraction of existing wealth, not necessarily in ensuring long-term sustainability. With limited state capacity to enforce laws, TNCs act with impunity. Following World Bank advice in 2002, the government of the DRC imposed a moratorium on new logging contracts. Currently 45 million ha is under contract. The government also rescinded contracts held by noncompliant concessionaires—roughly 25 million ha. According to the Bank, this was an "unprecedented move." However, it soon met with resistance:

> "In 2003, interest groups began to undermine the reform strategy by taking advantage of the fragile coalition government. Reportedly, six million hectares of contracts were allocated or exchanged in spite of the moratorium, and another 2.4 million hectares that had been cancelled in 2002 were reinstated. The rules for the legal review were altered several times in successive ministerial decrees (July and November 2004, June 2005) and, contrary to what had been agreed, the legal review did not start." (World Bank, *Questions and Answers—World Bank support to sustainable management of forests in the Democratic Republic of Congo* at http://web.worldbank.org).

[5] For example, of the top 10 companies, nine are European (five French, three Italian, one Belgian, one Lebanese). Together they are logging more than half of the forest concessions (2,035,000 ha of 4,054,000 ha).

Thus, despite growing international commitment to help African states turn forest resources toward national development, powerful vested interests—both national and foreign—within the state continue to practice resource capture for their own narrow ends.

4.3 Mystic Space

A third significant motivating factor behind sustainable conservation and management strategies is the place of forests in local and global imaginaries. Often the narratives which emerge are antagonistic, particularly where global actors (e.g., WWF, National Geographic Society) tend to conceptualize rainforests as wild, uninhabited places and local actors tend to regard them as varyingly religious, threatening, lived space, and important resource components in livelihood strategies (Alexander *et al.*, 2000). Reconciling these often contrasting perspectives is not easy. However, it is, I would argue, extremely important for global actors to move away from the pristine forest narrative if sustainable and reasonable approaches to forest conservation and management are to be achieved (for some ideas, see Fairhead & Leach, 2000).

World Resources Institute constructions illustrate the dangers inherent in conceptualizing the forest as untraversed space. In a recent publication, WRI uses two maps placed side-by-side to demonstrate the fragmentation of forests in the Congo River basin. These illustrations show the basin without (i.e., before) and with (i.e., after) roads. In their words: "road networks provide an unambiguous and easily measured, if conservative, indicator of the extent of human-induced fragmentation" (WRI, 2000, p. 94). The before picture is, quite understandably, an unbroken sea of green. The after picture is, equally understandably, replete with blue and light brown scars,[6] most notably along the banks of the Congo River. However, to assume that prior to the modern era there was only closed canopy is to render invisible the people who have historically used the forest and environs as lived space. The before picture, therefore, cannot be considered anything more than an artist's rendering; it has no basis in fact. The after picture, while accurate, is misinterpreted: what is absent is an understanding of the Congo River as the road network itself. The area deforested by infrastructure simply traces the long-established pattern of human settlement along the banks of the Congo River. Thus, people must be placed back into the rainforest picture. To fail to do so is to repeat the mistakes of the past, where conservation equaled dispossession and exclusion and led to widespread alienation from local people (see for example Schroeder, 2000).

To some extent, this is being done—particularly at regional level. For example, in the DRC's 2002 Forest Code which sets out basic principles for sound forest management, there is clear commitment to involving local people in resource use decisions and management. As highlighted by the World Bank, the major innovations of the forest

[6] Green shaded areas connote unbroken forest blocks greater than 10,000 km^2. Blue shaded areas—denoting by far the most dramatic change from the "before" illustration—connote unbroken forested areas of 1,000–9,999 km^2. Light brown areas connote 10–999 km^2 of unbroken forest cover.

code can be summarized as follows (see, World Bank, *Questions and Answers—World Bank support to sustainable management of forests in the Democratic Republic of Congo* at http://web.worldbank.org):

- Mandatory implementation in all production forests of sustainable management plans including the protection of biodiversity, (art. 71, 99, 100).
- Preservation in all production forests of traditional users' rights, including those of indigenous people (art. 44).
- Transferal of 40% of forest area fees to decentralized administrative entities ("*Provinces*" and "*Territoires*," art. 122).
- Mandatory in-kind contributions by forest companies to rural development in neighboring villages ("*Cahiers des Charges*," art. 89).
- Direct management of forest by local communities ("*Concessions des Communautés Locales*," art. 22);
- Establishment of national and provincial forest consultative councils; various types of protected areas to total of 15% of the national territory (art. 14).
- Allocation of future logging rights through a transparent process following consultations with local communities (art. 83, 15, 84).
- Provisions for new nonextractive uses of forests and the valorization of environmental services (art. 72, 87, 96, 119).

The vast majority of international environmental NGOs continue to frame the rainforest as a pristine wilderness to be preserved, contrasting rapacious rebels and refugees with peaceful charismatic mega-fauna such as mountain gorillas (Chapin, 2004). While this may be a rhetorical ploy, such framings lead to particular policy interventions that may in fact hinder peace processes and conservation efforts rather than help them.

5 MANAGEMENT STRATEGIES: SUCCESSES AND FAILURES

5.1 Mixing Global and Local

Approaches to sustainable forest management are growing increasingly sophisticated. Whereas bilateral initiatives still exist, most forest-centered economic, environmental and governance activities are multilateral in character.

In terms of bilateral, forest-specific initiatives, Canada, as a self-described world leader in the forestry industry, for example, has extended its domestic program of "working model forests" to the international level. The government did so in the wake of the 1992 Earth Summit. These model forests act as living examples of the possibilities of sustainable conservation and utilization. At the same time, Canada is working with the government of Cameroon to improve its capacity to "implement its forestry policy in the context of responsible environmental management" (see www.acdi-cida.gc.ca). At present there are two multiyear projects concentrating on the southern zone of Cameroon's forests. The stated aim of the Cameroon Environmental Protection Support project is to see that "local NGOs are trained and active with government" and that "forest administration works in harmony with communities and the environment" (*Ibid.*). The second project, Sustainable Management of Cameroon Forests, seeks to support the forest administration with its reform process and to support Cameroon's implementation of forestry policy "in a context of responsible environmental management" (*Ibid.*).

Most OECD countries are involved in myriad multilateral initiatives that bring together research centers, private sector companies, civil society groups, community based organizations, IGOs, IFIs and states so turning discrete geographical spaces into crucibles of global political economy. These activities are set within the broad context of the Forest Principles initially hammered out at Rio whose holistic focus was reiterated at the United Nations Forum on Forests (UNFF) Ministerial Declaration and Message to the WSSD. Each of these actors brings different capabilities to programs and projects. Research centers such as CIFOR bring information to inform policy. Civil society groups act as watchdogs, independent overseers, and lobbyists whose global campaigns of naming and shaming corporate bad actors are increasingly sophisticated and effective. Donors assist with sectoral reform (based on transparency and rule of law), establishing market opportunities and guidelines (e.g., chain-of-custody from stump to export, voluntary compliance), developing smart partnerships, and building financial, infrastructural and human resource capacity. For Collier (2004, p. 11), actors need to work toward "effective global governance of the markets that have come to facilitate rebellion and corrupt governance ... While globalization provides rebels with new opportunities, it also makes them more vulnerable to international pressure ... provided the international community is willing to exert it."

In conflict settings such as Liberia and the DRC, forest-centered initiatives are buttressed by legal instruments such as sanctions (against blood diamonds, illegal logging, arms sales), asset freezing and restrictions on the movement of belligerents, the dispatch of UN forces (under the rubric of the Responsibility to Protect, for example) as peacemaking, keeping, and building missions, and the use of track two and three diplomacy to work toward the end of violent conflict and to minimize humanitarian disasters.

Both bilateral and multilateral approaches to forest resource management utilize integrated natural resource management approaches. For example, at the multilateral level is the USAID-supported CARPE, itself one element within a wider Congo Basin Forestry Program. CARPE draws together a wide variety of actors in the effort to preserve remnant blocks of closed canopy habitat. Those involved include the following US-based groups: African Wildlife Foundation, Biodiversity Support Programme, Conservation International, Peace Corps, Innovative Resource Management, World Conservation Society, World Resources Institute (WRI), World Wildlife Fund, the US Fish and Wildlife Service, the US Department of Agriculture Forest Service, the University of Maryland's Department of Geography, and USAID. In addition, there are ten local African partners and two multilateral actors, the FAO and the World Bank. The FAO role is to develop a data system; the World Bank seeks to create a demand-driven environmental information database. Under the rubric of CARPE, the WRI, for example, has developed its Frontier Forests Initiative, drawing into its fold a number of other partners, among them the National Geographic Society.

While many of these initiatives are well-meaning and important—especially those which seek to gather and make readily available accurate data—the majority of approaches to managing and sustaining Africa's forests are, in my estimation, both self-serving and somewhat naïve. Many of these initiatives seem preparatory: when and if there is peace, we will be ready to implement these policies. While this may seem reasonable to many, the next section seeks to demonstrate the folly of much of this activity.

6 CONCLUSIONS: SEEING THE FOREST FOR THE TREES

This chapter has described the fluid and widespread nature of violent conflict on the African continent. This conflict is primarily subnational or transnational (but not specifically interstate), diffuse and persistent. Fatalities result less from direct violence than they do from structural violence set in place by historical social relations of production made worse by post-Cold War changes to state and regional forms of authority and legitimacy (Clapham, 1996; Ellis, 1999). Characterizing Africa's forest wars as bizarre and barbaric deflects attention from the clear fact these are global conflicts conducted across a particular geographical space for many different reasons.

While forest degradation and violent conflict are linked, both of these processes have persisted through time and are neither mutually dependent nor vary directly. To suggest that they are, for instance through a resource curse narrative, ignores both African and wider world histories. Forests and forest resources have come to play a central part in post-Cold War African political economies. As formal state forms of production and authority have collapsed, shadow economies and medieval-style multiple centers of power have emerged. Global actors—from arms merchants to logging companies, from diamond dealers to *Medecins Sans Frontiers*—have reacted to the many opportunities that these new structures present. In my view, two factors facilitate the persistence of these negative processes of deforestation and violent conflict. One is the African state form; the other is the neoliberal world order. Unless these two inter-related factors are addressed, environmental degradation and violent conflict will persist.

6.1 Simulating the Westphalian State Form

The second opening epigram to this chapter from the UNDP argues that global security is dependent on strong states and strong interstate cooperation. Many observers argue that African states are weak due to their own bad governance (World Bank, 1989; 1997). While not disputing the contribution of African governments and structures of governance to current difficulties, one must remember that Africa's states, like institutions everywhere, reflect the needs of the past, not those of the present, let alone hope for the future (Dovers, 2000). These needs were born of colonial and imperial exploitation. African leaders inherited states constructed to facilitate the extraction of wealth (slaves, minerals, animals, timber, food and cash crops) from the continent. Belgian, Portuguese and French forms of colonialism were especially pernicious in this regard (for example, see Hochschild, 1999).

Far from being any more barbaric and savage than violent conflicts anywhere else in the world, the civil wars in Liberia and Sierra Leone are eminently understandable within the contexts of historical state formation in Africa and contemporary, post-Cold War economic decline (see, for example, Clapham, 1996; Ellis, 1999, Reno, 1999). Liberia, for example, was formed as a free territory for repatriated American slaves. These ex-slaves and their descendents came to rule a territory they rarely set foot in. With time, Liberia came to resemble a classic colony in the sense that a small core of individuals occupying a narrow strip of coastline dominated a large geographical hinterland and its many peoples. By the mid-1920s, Liberia embodied the patronage politics of a *rentier* state. In 1926, "the American rubber tycoon Harvey S. Firestone Sr., with the support of the US State Department, engineered a major business coup by signing a 99-year lease on a million acres of land in Liberia, soon to become the world's largest rubber plantation" (Ellis, 1999, p. 44). It was from this disaffected hinterland—this home turned plantation—that Charles Taylor came to draw his army. Taylor himself was born in Arthington, just northeast of the coastal capital, Monrovia, and was schooled in the U.S.

It is no surprise to say that Taylor was not the first of his kind either in Liberia or in Africa; nor will he be the last. As long as center–periphery style state forms persist, there will always be fertile ground for revolt from the margins. Moreover, given African states' general inability to penetrate much beyond the capital city (Herbst, 2000), the hinterland will regularly give rise to its own forms of power and authority. The Forces Nouvelles in northern Côte d'Ivoire are only the latest in a long line of such revolutionary movements. To survive in a post-Cold War era, rebel groups "irrespective of the motivation for the rebellion . . . must also become a business organization" (Collier, 2004, p. 3). Where easily exploitable, highly portable and globally valuable resources such as alluvial diamonds are available, it seems not unreasonable to conclude that central states may never gain control of their hinterlands (Parker *et al.*, 2004). The presence of ideologues such as Libya's Gaddafi only adds to the unstable mix.

It is no accident that the zones of violent conflict on the continent tend to correspond to those areas dominated by these three former colonial masters. These states have been variously described as weak, predatory and patrimonial—all quite accurate in my estimation (see Bayart, 1995). Policies which work through these states overestimate their capacity to ever approximate their Western counterparts—they are fundamentally different; they were created for fundamentally different purposes. In many ways, then, reform is simulation, rather than real reform and it will last only so long as donor funds are available. Moreover, as Ferguson (1990) so aptly demonstrated in the case of World Bank projects in Lesotho, local actors operate in local contexts. This means that any funds or other forms of support from outside the local will be twisted to fit the local context. Projects therefore do not fail—they achieve very specific locally-determined ends. These are ends which may be very different from those desired by donors, however.

6.2 Liberal Peace Liberal War

Duffield (2001) argues that the attempted construction of a neoliberal world order has given form to very specific kinds of warfare, a liberal war where globalization and structural adjustment have combined to undermine existing social orders within states narrowly articulated with global capitalism. Whatever positive activities are undertaken in the name of resource conservation, they are ultimately undermined by economic structural adjustment conditionalities and weak state debt burdens. Among other factors, country-specific liberalization and deregulation have led to a decline in the global price of timber products, as producers flood world markets (FAO, 2003). Privatization has further undermined authority and legitimacy in highly-contested state forms, so encouraging the rise of warlordism (Greenpeace International, 2005a).

While the political cross conditionality of first-past-the-post elections may satisfy donor criteria for openness, it simply rearranges the power political deck chairs once every 4 years so leading to a cycle of kleptocracy and retribution. As a sign of good faith to citizens and donors alike, Liberia's President Johnson–Sirleaf sacked her entire Ministry of Finance in a high-profile crackdown on corruption. Observers would be wise to remember, however, that similar actions were taken in Kenya, Malawi and Zambia following deposition of dictators—all countries currently embroiled in comprehensive corruption scandals. Greenpeace (2005b), for one, has begun to explore the links between World Trade Organization rules and forest loss; but formal donor policies are all underpinned by firm commitments to neoliberal globalization. Given African weak state forms, neoliberal globalization fosters resource capture rather than creates the basis for sustainable resource management (Homer-Dixon, 1999). Put differently, Western actors deliberately focus on the fragile and stunted African state tree, without acknowledging that it exists in a wider forest of economic globalization.

6.3 Policy Implications

There are a number of very clear policy implications to arise from this overview and critique. I will, however, limit myself to three of what I consider to be the most important.

First, there is an abiding difficulty particularly in the social and policy sciences with reconciling macro- frameworks with micro- realities. While it may be essential to contemplate global governance of the world's "lungs," attempts to prevent exploitation of these resources by local peoples without addressing the underlying basis of the poverty and desperation that drives some of them to ransack their own source of livelihood will fail (Dalby, 2002). Similarly, to assume that one policy fits all—i.e., economic structural adjustment and multiparty elections—is to ignore the realities of people on the ground. Despite policy pronouncements regarding local stakeholder participation, transparency and accountability, there is no empirical evidence to suggest willingness by governments in West and Central Africa to relinquish power, either over an essential resource or over people. As recent experience with attempts at water reforms and community based natural resource management programs in Southern Africa show, powerful people tend to manipulate any intervention from outside to their own advantage (Alexander

& McGregor, 2000; Twyman, 1998; Swatuk, 2005). In the context of shrinking formal economies and often disintegrating polities, this is to be expected.

Those truly interested in conservation alone would do well to recognize that state makers, rebels and warlords only understand the hard bargaining of threat and compromise. No amount of soft talk about reform and democratization will change this fact. Saving the rainforest may require some serious trade-offs along the lines of the 1980s debt/environment swaps—irrespective of the regime in power and its behavior.[7] Placed in this context, recent Conservation International ideas to buy closed canopy forest seem rational rather than bizarre. Accompanying this must be strong OECD pressure for formal regulation of forestry TNCs—such companies will only respond to concerted international legal action. But as long as countries such as France and Italy resist compliance (as a market for conflict timber) or enforcement (of their own TNCs), the free for all across African forests will continue.

Second, sustainable forest management is not a sectoral activity that is fungible in global space and time. Rather, it is a site-specific developmental activity that seeks either to change the way people live their lives, or to assist them to maintain their current lifestyles while adopting more sustainable practices. Both ways are highly political acts and resistance—either from the communities in question or the central state or both—should be expected. Understanding the context for sustainable forestry management may therefore require anthropologists, not economists.

In this regard, donor and multilateral activities focusing on community forestry management projects in concert with land reform and more productive forms of food production are to be encouraged. In a difficult socio-political situation, however, bypassing the state to work directly with communities may be problematic both for the donor and for the local community. Yet donors have dollar power which should be exercised with caution. Unfortunately, the overwhelming preoccupation with structural adjustment programs and possible debt forgiveness in exchange for harsh political and economic reforms is the wrong kind of dollar power. It is no accident that those states most readily pursuing structural adjustment programs—e.g., Cameroon and Ghana—have witnessed the most widespread degradation to their forests. Liberalization in the context of debt destroys forests; it does not preserve them. Sustainable forest management may thus begin with debt forgiveness, strong market regulation, and increased access for African textiles and cash crops to Western markets.

Lastly, the weak African state form, articulated marginally into global capitalism,[8] and dominated by particular ethnic and class groups encourages an ongoing cycle of

---

[7] If you balked at this point, it should then give you pause to consider both your own motives and the consequences of your actions. West and Central African state-society relations are highly volatile. Relations between "governors" and "governed" are by no means settled or peaceful. To interact with and enact policies through elites will have very specific effects at mass level. How important is conservation for its own sake in the presence of such social relations? Indeed, is "conservation" here anything more than a ritualized enactment of what you would like it to be, rather than what it in fact is?

[8] Bannon and Collier's work on resource dependence and the need for economic diversification is important in this regard: enclave and/or rentier economies will become more unstable in the absence of deepening capitalist development and job creation.

resource degradation (be it formal, such as oil in Angola and Nigeria or informal such as coltan in the eastern DRC) and political instability. For Homer-Dixon and Blitt (1998, pp. 1–2), environmental scarcity is a consequence of the depletion of renewable resources such as forests, soil, lake and river water and the stratospheric ozone layer. They provide both goods (fish, plants, water) and services (fertile soil within which plants root and grow, carbon stores, pollution filters), with numerous linkages.

Scarcity, Homer-Dixon states, can come about in one of three ways: as a result of increased demand (demand-induced); as a result of decreased supply (supply-induced); or as a result of unequal access to and distribution of these resources (structural). These push and pull factors are said to often coexist. In the event of scarcity, two processes are set under way. Resource capture takes place by those with the means to do so, while those without are ecologically marginalized.

Resource capture and ecological marginalization are rife throughout the West/Central African landscape. The rainforests are part of this landscape, as sites of struggle and sources of friction, as providers of both global and local goods and services, as a thing and a social process, at once material and ever-changing. To "save" the rainforest begins with putting it back into the landscape, and locating this particular landscape in wider, equally complex and confounding constructed realities. Extracting "deforestation" from these wider processes and structures and relocating it within a narrow—and narrowly state-centric—discourse of violent conflict and environmental scarcity, however, is a temptation to be resisted. Acknowledging and articulating the interrelationship between violent conflicts, resource degradation, weak states and neoliberal globalization is a step in the right direction. Greenpeace, among others, has taken this step; OECD donor states have not.

REFERENCES

Alexander, J., McGregor, J., & Ranger, T. (2000). *Violence and Memory: One Hundred Years in the 'Dark Forests' of Matabeleland.* London: James Currey.

Alexander, J., & McGregor, J. (2000). Wildlife and politics: CAMPFIRE in Zimbabwe. *Development and Change, 31*, 3.

Bannon, I., & Collier, P. (Eds.) (2003). *Natural Resources and Violent Conflict: Options and Actions*. Washington, DC: World Bank.

Bayart, J.-F. (1995). *The State in Africa: The Politics of the Belly*. London: Longman.

Blundell, A. G. (2005). Conflict timber and Liberia's War. *ETFRN News*, 43–44/05.

Chapin, M. (2004, Nov/Dec). A challenge to conservationists. *Worldwatch*.

Clapham, C. (1996). *Africa and the International System.* Cambridge: Cambridge University Press.

Collier, P. (2004). *Reducing the Global Incidence of Civil War: A Discussion of the Available policy instruments*, from www.mandela.inwent.org/ef/military/collier.htm.

Cox, R. W. (1996). *The New Realism*. New York: Palgrave.

Dalby, S. (2002). *Environmental Security*. Minneapolis: University of Minnesota Press.

Dovers, S. (2001). Institutional barriers and opportunities: Processes and arrangements for natural resources management in Australia. *Water Science and Technology, 43*, 9.

Duffield, M. (2001). *Global Governance and the New Wars,* London and New York: Zed Books.

Ellis, S. (1999). *The Mask of Anarchy: the Destruction of Liberia and the Religious Dimension of an African Civil War.* New York: New York University Press.

Fairhead, J., & Leach, M. (2000). Reproducing locality: A critical exploration of the relationship between natural science, social science, and policy in West African ecological problems. In V. Broch-Due & R. A. Schroeder (Eds.), *Producing Nature and Poverty in Africa*. Stockholm: Elanders Gotab for NAI.

FAO. (2003). *Forestry Outlook Study for Africa: Subregional Report. Central Africa*. Rome: FAO.

Ferguson, J. (1990). *The Anti-Politics Machine: Development, Depoliticization and Bureaucratic Power in Lesotho*. Cambridge: Cambridge University Press.

Global Witness. (2001, Sept). *Taylor-made: The Pivotal Role of Liberia's Forests and Flag of Convenience in Regional Conflict*.

Global Witness. (2004, Dec). *Dangerous Liaisons: The Continued Relationship between Liberia's Natural Resource Industries, Arms Trafficking and Regional Insecurity*.

Global Witness. (2005, June). *Tinker Taylor Soldier Spy: Uncontrolled Resource Exploitation, Charles Taylor's Manipulation and the Re-recruitment of Ex-combatants are Threatening Regional Peace*.

Greenpeace International. (2005a, Dec). *Illegal Logging in Cameroon: How French Government Action is Fuelling Rainforest Destruction*.

Greenpeace International. (2005b, Dec). *Trading Away our Last Ancient Forests: The Threats to Forests from Trade Liberalization under the WTO*.

Herbst, J. (2000). *States and Power in Africa: Comparative Lessons in Authority and Control*. Princeton: Princeton University Press.

Hochschild, A. (1999). *King Leopold's Ghost*. Boston and New York: Mariner Books.

Homer-Dixon, T. (1999). *Environment, Security and Violence*. Princeton: Princeton University Press.

Homer-Dixon, T., & Blitt J. (Eds.) (1998). *Ecoviolence: Links among Population, Environment and Security*. Boston: Rowman and Littlefield.

Johnson, P., & Chenje, M. (1994). *State of the Environment in Southern Africa*. Harare: Zed Press.

Parker, M., Feil, M., & Kramer, A. (2004). *Environment, Development and Sustainable Peace: Finding Paths to Environmental Peacemaking*. Wilton Park Paper, November. Wiston House: Sussex, U.K.

Raskin, P., & Lazarus, M. (1991). Regional energy development in Southern Africa. *Annual Review of Energy and Environment, 16*, 145–178.

Reno, W. (1999). *Warlord Politics and African States*. Boulder, CO: Lynne Rienner.

Richards, P. (2005). The Mano River conflicts as forest wars. *ETRFN News*, 43-44/05.

Schroeder, R. A. (2000. Producing nature and poverty in Africa: Continuity and change. In V. Broch-Due & R. A. Schroeder (Eds.), *Producing Nature and Poverty in Africa*. Stockholm: Elanders Gotab for NAI.

Shambaugh, J., Oglethorpe, J., Ham, R. (with contributions from Sylvia Tognetti). (2001). *The Trampled Grass: Mitigating the Impacts of Armed Conflict on the Environment*. Washington, DC: Biodiversity Support Program.

SIPRI Yearbook. (2000). Stockholm International Peace Research Institute, Stockholm.

SIPRI Yearbook. (2005). Stockholm International Peace Research Institute, Stockholm.

Swatuk, L. A. (2005). From 'Project' to 'Context': Community based natural resource management in Botswana. *Global Environmental Politics, 5*(3), 285–323.

Tevera, D., & Moyo, S. (Eds.) (2000). *Environmental Security in Southern Africa*. Harare: SAPES Books.

Twyman, C. (1998). Rethinking community resource management: Managing resources or managing people in western Botswana. *Third World Quarterly, 19*, 4.
United Nations Development Program (UNDP). (2000). *Human Development Report 2000.* Oxford: Oxford University Press.
United Nations Development Program (UNDP). (2005). In *Larger Freedom: Toward Development, Security and Human Rights for All.* New York: United Nations.
United Nations Environment Program (UNEP). (1999). *Global Environment Outlook 2000.* London: Earthscan.
USAID. (2005, Aug). *Foreign Assistance Act, Section 118: Tropical Forests FY 2004: A USAID Report to Congress.* Washington, DC: USAID.
Wolfire, D., Brumner, J., & Sizer, N. (1998). *Forests and the DRC: Opportunities in a Time of Crisis.* Washington, DC: World Resources Institute.
World Bank. (1989). *Sub-Saharan Africa: From Crisis to Sustainable Development.* Washington, DC: The World Bank.
World Bank. (1997). *World Development Report.* New York: Oxford University Press.
World Resources Institute (WRI). (2000). *World Resources 2000–2001.* Washington, DC.

CHAPTER 7

CONFRONTING CONFLICT TIMBER

STEVEN PRICE, DEANNA DONOVAN, AND WIL DE JONG

1 INTRODUCTION[1]

"Conflict timber" is a term that has only come into use recently, yet the broader phenomenon—the financing of violent conflict through the exploitation of natural resources—has a long history. For the purposes of this study, conflict timber is defined as wood that has been traded or taxed at some point in the chain of custody by armed groups, be they rebel factions or state militaries, or by a civilian administration involved in armed conflict[2] to finance hostilities or otherwise perpetuate conflict.[3] The connection between the timber trade and the finance of war has garnered international attention recently because of its role in several prolonged and destructive conflicts in Asia and Africa. Timber has most prominently figured in conflicts that have affected Liberia, the Democratic Republic of Congo (DRC), and Cambodia, as described in the chapter by Le Billon (Chapter 2), but it has also played a role in conflicts in Myanmar (Burma), Guinea, and Sierra Leone.[4] Here we examine the basic characteristics of the conflict timber phenomenon and review options available for preventing and controlling its occurrence.

Although conflict timber is not a widespread problem in timber producing nations or in developing nations affected by armed conflict, governments have begun to recognize that the implementation of policy responses to prevent or diminish the trade of

[1] The chapter is based on a CIFOR report: Conflict timber. The World Bank's Forest Governance Program provided financial support for preparing the report. The findings, interpretations, judgments and conclusions expressed in that report, and this chapter are those of the authors and should not be attributed to the World Bank, to its affiliated organizations, or the members of the Board of Executive Directors or the governments they represent. Global Witness worked with CIFOR to provide information for the case studies outlined in the report.

[2] Violent conflict occurs on many different scales. In this study, conflict is used synonymously with war, armed conflict and violent conflict in which at least 1,000 deaths have resulted.

[3] This definition is a modified version of the definition used by the British nongovernmental organization, Global Witness (2002a).

[4] In these wars, the control of timber or the proceeds of its commerce has not incited or sparked the conflicts, but rather it has played an influential role in their evolution.

W. de Jong, et al. (eds.), Extreme Conflict and Tropical Forests, 117–132.

conflict timber can serve many development, democracy, and governance objectives. Addressing the factors that facilitate the exploitation and trade of conflict timber offers an opportunity to contribute to achieving the goals of numerous regional and global agreements and declarations on arms control, development and resource management.

Conflict timber is distinguished from other forms of legal and illegal timber trade by its role in funding arms proliferation and perpetuating violence and socio-political upheaval. However, conflict timber shares many characteristics with the broader phenomenon of illegal logging.[5] Much of the conflict timber trade is in fact illegal and many of the same companies, trade networks, financial mechanisms and entrepreneurial methods are involved. Perhaps most significantly, both conflict timber and illegal logging flourish where governance is weak, financial controls are insufficient, and the participation of civil society in resource management is limited. Because of these common features, some international measures undertaken to combat corruption and illegal and unsustainable logging may prove useful in addressing the trade in conflict timber and vice versa.

Individual nations and several intergovernmental bodies have recognized the conflict timber problem and begun to take steps towards addressing it, but action taken to date has been piecemeal. The United Nations (UN) significantly raised the international profile of the role of timber in African conflicts through the publication of reports by expert panels commissioned by the Security Council and finally by banning the international trade of Liberian timber in 2003. Member states of the G-8 declared their willingness to work "with African governments, civil society and others to address the linkage between armed conflict and the exploitation of natural resources" in the *Africa Action Plan*, drafted in 2002 (Le Billon, 2002). The Commission of the European Communities' proposed Action Plan for Forest Law Enforcement, Law and Governance, drafted in May 2003, specifically recognizes conflict timber and pledges in general terms to address the problem through its cooperation programs.

There remains a need to develop mechanisms for financial controls and oversight and systems to track the flow of timber and other so-called conflict commodities from conflict zones to consumer countries. The emerging international responses to the trade of diamonds in regions affected by conflict have shown that it is necessary to address the trade of not just one but a variety of resources that are used to finance armed conflict. In fact in most cases, timber is not the primary or sole commercial interest of insurgent groups, militias and militaries. This means that restricting the trade of one commodity is unlikely to affect the course of conflict if armed groups have access to lucrative, unregulated alternatives.

Significantly reducing the use of natural resource revenues in armed conflicts will require a mix of short-term and long-term efforts. The dramatic pace of resource

---

[5] Illegal logging is the harvest, processing, transportation, purchase, or sale of timber in violation of national laws. The illegality of forest practices varies by countries, but they may include, obtaining forest concessions through bribery or other forms of corruption, the removal of timber without permission, the cutting of protected species or the removal of under or oversized trees, the failure to pay fees or taxes, among other practices.

exploitation, military campaigns and humanitarian emergencies may require the implementation of measures with immediate effect—including UN bans on the trade of timber and other sanctions and controls. Over the long term, sustained efforts and initiatives to improve governance in countries with significant stocks of high-value tropical timber will be needed to address the conditions and contexts that enable governments and armed groups to exploit conflict timber. The countries whose governments, companies, or citizens are involved in financing, transporting, marketing and consuming conflict timber will need to take a leadership role in developing such initiatives as well as implementing effective regulatory and market mechanisms.

2 THE PRINCIPAL ASPECTS OF CONFLICT TIMBER

The problem of conflict timber, although unique in every country, can be characterized according to three aspects: Its institutional context, the main participants, and their motives. Understanding these elements is essential to understanding how to approach this problem.

2.1 The Institutional Context

Poor governance is perhaps the most common feature or condition found in the countries where conflict timber has emerged as a serious problem. In broad terms, governance factors set the rules and institutional arrangements that essentially determine who has access to forest resources, what happens to the benefits of resource exploitation, and how commodity markets—including clandestine networks—develop and are controlled. Some of the most relevant governance problems associated with the conflict timber phenomenon include high levels of corruption, the ineffective or arbitrary rule of law, a lack of accountability for military and state officials, and limited civil society involvement in natural resource control and decision making.

In the natural resource sectors, poor governance often means that there is little civil society oversight or public safeguards for ensuring the lawful and transparent use of the royalties, fees and taxes collected from forest concessionaires and other users. Under these conditions, funds from state-sanctioned or supervised timber exploitation can be diverted from the public purse to "extra-budgetary activities" (Smillie, 2002, p. 9; UNSC, 2001a). In the context of violent conflict, these may include the financing of paramilitary groups, weapons purchases, or key political supporters—particularly common when national political leadership is threatened by armed opposition.

One of the characteristics of the rule of law in the countries affected by conflict timber, is that forestry legislation is often inadequate, antiquated and poorly administered. Poor enforcement is due to many factors, including limited institutional capacity, corruption and the intimidation of regulatory agencies and judiciaries. Law enforcement is more challenging where public authority is violently contested and states have little capacity to control their own territory or what happens in it. Sporadic violence or low intensity conflicts may be sufficient to disrupt or suspend law enforcement, whereas full-scale war frequently results in the collapse or suspension of most government institutions and functions in contested areas—a situation that facilitates the looting of natural resources by security forces and armed groups (Global Witness, 2002b).

2.2 Motives and Main Players

In areas of armed conflict, the trade in conflict timber provides operational finance, political power and personal enrichment for private companies, government regimes, state militaries, rebel armies and private militias. Revenues from the sale, transport, and taxation of timber have enabled state militaries, insurgent groups and private militias to purchase weapons, recruit and reward combatants, forge strategic commercial and military alliances and enrich their leaders.

Once violence has been initiated and combatant forces are mobilized, it is the financial viability of the militaries or rebel groups that largely determines the duration of violent conflict (Collier *et al.*, 2001). Though timber is rarely the principal or preferred commodity traded by armed groups and militaries, it can offer combatants a relatively rapid and straightforward means for establishing an income flow and amassing a significant amount of wealth. Compared to petroleum, the exploitation and processing of timber requires relatively low levels of capital investment, technology, and skill, and therefore can attract a wide range of firms and operators (Thompson & Kanaan, 2004).

The potential profits to be earned from conflict timber and other natural resources found in forested areas can be a powerful incentive for militaries and armed groups to extend their influence over forestlands. Where high-value timber exists over large tracts of forest extending across political boundaries, multiple political and military contenders may simultaneously vie for and benefit from the same resource. For example, in Cambodia during the 1990s, multimillion dollar timber revenues financed arms procurement by both the Royal Cambodian Armed Forces and their enemy, the Khmer Rouge. In the business of conflict timber, the political and ideological distinctions of the various client groups may therefore be blurred by a common commercial purpose.

Conflict timber does not automatically emerge as a factor, however, in all timber-rich countries that experience civil war or international conflict. The degree to which militaries or armed groups exploit and trade timber depends on the geography of territorial control by warring groups, the conditions of transportation infrastructure, the availability of productive capacity and technology, market access, and other political and logistical issues. Some of these factors explain the variable levels of logging that have been witnessed across the DRC's different forested regions during wartime. Despite the existence of vast stocks of valuable timber in areas controlled by rebels, large-scale timber harvesting and export was often inhibited by limited infrastructure and market access.

2.3 Governments and State Militaries

In most cases, conflict timber occurs with the complicity or direct involvement of state actors. Unlike other more compact and lightweight conflict commodities, such as diamonds or gold, the exploitation and transport of timber are not easily concealed and securing access to markets for this bulky commodity usually requires the consent of state organizations. This means that conflict timber is almost always logged and traded with involvement or assistance of government authorities. Organizations aligned with or controlled by the State, most frequently state security forces or government-backed logging companies, are therefore more likely to be involved in the conflict timber trade than rebel groups (Thomson & Kanaan, 2004).

States may justify military involvement in the forestry sector on the grounds of military necessity or strategic concerns. The participation of the military in the forestry sector ranges from taxation of forest production to ownership of forestry enterprises and control of forestlands. Collaboration between militaries and the private sector may involve a merging of commercial logging and military interests.[6] In desperation, states may sell or barter forest concessions to foreign companies or governments in exchange for capital, military equipment, or other strategic support. For example, the government in the DRC sold natural resource assets, including logging concessions reportedly covering 33 million hectares, to secure foreign support while foreign occupiers engaged in the looting of natural resource stockpiles and systematic exploitation of mines and forests.[7]

2.4 *Private Enterprises*

A wide range of business enterprises are often involved in the trade of conflict timber. They include companies of various sizes that operate along the two-way commodity chain that stretches from the forest to consumers overseas. Some of the most conspicuous roles in the trade include purchasing timber from combatants or their known representatives, trading arms financed by timber sales, harvesting timber with illegally or corruptly obtained permits, laundering funds from the sale of conflict timber, processing or shipping conflict timber, and organizing militias to control logging areas and control local populations.

The logging companies that are attracted to the conflict timber trade are often opportunistic and unscrupulous enterprises that specialize in rapid resource extraction in the context of failing or failed states. Such firms frequently must ally themselves with one or more armed factions in order to access forest resources in areas affected by conflict. To compensate for their special privileges, these firms often make extra-contract payments to relevant state or nonstate actors, arrange shipments of arms or other military supplies, or transport materials and combatants into remote areas.

---

[6] The joint venture formed between the Zimbabwean military-controlled company OSLEG (Operation Sovereign Legitimacy) and the DRC-based Comiex is an example of how the timber trade and military authority become combined within a corporation (see Global Witness, 2002a).

[7] It was sufficient to finance a significant part of the operations of the Rwandan and Ugandan armies. Zimbabwe's less commercially successful military intervention in the DRC was also motivated in part by the promise of wealth from resource exploitation (see UNSC, 2001a).

In Cambodia and Liberia, private logging companies have mobilized their own armed militias, in effect militarizing their control and exploitation of forest concessions (Global Witness, 2002b). These company militias have been used to assert control over local populations, discourage protest against logging, and forcibly encroach on communal lands. In Liberia, they are alleged to have intimidated local communities, destroyed and stolen goods, and expropriated farms and forest land (SAMFU, 2002). Logging company militias also reportedly provided security services to the Liberian government, policing forest areas and even engaging in combat against the rebel insurgents, namely the Liberians United for Reconciliation and Democracy (LURD) (Global Witness, 2002c).

2.5 *Rebel Groups*

Forests have long served as refuge and operational base for rebel armies, providing insurgents with food, fuel, building materials, and sources of finance. Rebels have derived income from forest resources by participating in logging operations, taxing or extorting funds from logging companies and timber traders, selling concessions in forest areas under their control, and looting logs and sawn timber. By participating in the trade of licit and illicit commodities many groups have increased their financial independence, thus reducing their reliance on and accountability to the rural communities from which they traditionally have drawn recruits, materials and other support. The decline in patronage of armed movements from foreign states following the end of the Cold War may have spurred increased self-reliance and the criminal entrepreneurial orientation of some rebel armies operating in forested regions. Armed groups are often drawn into forests regions in part because timber and arms often flow through the same trade networks.[8] For these groups, these factors can make the control of forests and local trade networks a strategic objective.

3 THE IMPACTS OF CONFLICT TIMBER

The trade of conflict timber has contributed to prolonging conflicts that have resulted in significant loss of life, incalculable suffering, widespread displacement of civilian populations, and degradation of globally significant forest reserves and protected areas. Disruptions to traditional livelihood activities, closure of legitimate economic enterprises, degradation of infrastructure, and loss of state revenues together produce serious negative consequences for the economies, societies, and governments of the affected countries. These conditions can complicate or simply prohibit the implementation of development initiatives or the kind of governance reforms that could address the underlying factors associated with the emergence of the conflict timber trade. It

[8] Evidence suggests that in Liberia, some enterprises within the timber industry have been the main channel for the illicit arms trade (UNSC, 2001a).

is important to note, however, that conflict timber is not the primary cause of violent conflict or the loss of human life, but rather is a contributing factor.

3.1 Social and Economic Impacts

Trade in conflict timber has both short and long-term implications for economic development and local livelihoods. The exploitation of forest resources by armed groups and commercial enterprises allied with state security forces often occurs in association with the expropriation of lands used or occupied by indigenous groups and other local populations, forcing people to abandon farms, forests, and businesses that traditionally have been their source of livelihood (Hart & Mwinyihali, 2001). The disruption of traditional productive activities has forced large segments of the populations in the war-torn states of sub-Saharan Africa to flee into neighboring forest areas where they have relied heavily on wild resources for food, fuel, building materials and income (Hart & Mwinyihali, 2001; Sato *et al.*, 2000).

The production and trade of conflict timber diverts the normal flow of timber from legitimate, established chains of production, resulting in the loss of income and earning potential for both local communities and governments. Conflict timber also undermines legitimate forestry enterprises by distorting market prices and unfairly competing with timber produced in well managed forests. The plunder of conflict timber and the accompanying growth of informal market networks can undermine the rule of law in the forest sector, particularly the exercise and enforcement of forest resource use rights, and thereby exacerbate existing natural resource governance problems. Together these factors are disincentives for long-term investment in forest management and can effectively undermine sustainable economic development.

3.2 Environmental Impacts

Because private and state-backed enterprises characteristically focus on removing the greatest volume of conflict timber in the shortest amount of time, logging is done with little regard for its social and environmental consequences. The on-site and off-site environmental impacts associated with the extraction of conflict timber therefore often include forest degradation, destruction of wildlife habitat, reduced hydrological function, and the loss of rare and endangered animal and plant species. If conflict timber logging significantly reduces the productive capacity of the forest or impairs the provision of important environmental services, it can negatively affect the long-term potential of forests to contribute to local development.

The impact of conflict timber on wildlife populations can be particularly severe where logging company militias and crews rely on hunting and the bushmeat trade as a source of food and/or income (SAMFU, 2002). By facilitating access to previously remote forests and their wildlife populations, the pursuit of conflict timber can facilitate the trade in bushmeat (Wilkie *et al.*, 2000). High levels of wartime poaching could conceivably limit the economic recovery of local communities that traditionally depend on forest wildlife. Furthermore, the loss of wildlife species in national parks and forest reserves diminishes the potential of nature-based tourism and thus future development options.

While the ecological impacts of conflict timber logging are clearly negative, the broader impact of violent conflict on forest biodiversity conservation, particularly in the short term, is somewhat ambiguous. Whether violent conflict in forested areas leads to an increase or decrease in overall logging levels depends greatly on the specific circumstances mentioned above. In some cases, violent conflict can alleviate pressure on forests by discouraging commercial logging or restricting agricultural expansion and settlement (McNeely, 2003). The rule of thumb is that if insecurity makes it difficult for companies to cut timber or transport wood to markets, logging will decline (Thompson & Kanaan, 2004). This occurred in parts of the DRC, where rebels controlled the country's richest primary forests but government forces controlled trade routes and market access. Likewise, the civil war in the Republic of Congo halted most logging operations in the south of the country during the late 1990s and national timber production dropped to its lowest level in 30 years (Forests Monitor, 2001). Indeed, in areas of deteriorating security and crumbling infrastructure logging may become logistically too challenging or financially too uncertain. In the longer term, given the ineffectiveness of state institutions and disorganization of civil society in the postwar period, any immediate benefits to forest conservation will likely be fleeting.

3.3 The Postconflict Period and Beyond

The need to restart national economies and rapidly organize economic opportunities for local populations in the postwar period can lead to increased pressures on forest resources. If the security situation and the investment climate are adequate, timber extraction can rebound as transportation infrastructure improves and unemployed people, including ex-combatants, seek a new livelihood. The distribution of timber use rights can be used as an enticement for the political settlement of conflict. Governments may attempt to appease or co-opt armed groups by permitting them to exploit timber in a certain area or to manage some aspect of the timber trade.

Because the institutional and regulatory environment is generally weak following the end of conflict, renewed logging activity can often outstrip the capacity of governments and local communities to reestablish resource management institutions and the rule of law. Under such conditions, the growth of natural resource markets can lead to increased levels of unregulated logging, and in worst case scenarios, the criminalization of the forestry sector (Le Billon, 2002). The postwar histories of Mozambique and the Republic of Congo exemplify how logging activity, temporarily depressed by war, can dramatically rebound in the postwar period with serious consequences for forests, wildlife and local communities (McNeely, 2003).

The strategic importance of controlling valuable natural resources and trade routes may grow in the emerging postwar order as ex-combatants and others actors attempt to enhance both their economic power and political position. The conflict in the DRC has revealed that wartime trade networks and logging by armed factions can persist into the postwar period, especially where conflicts have a ragged ending. When foreign armies pulled out of the country, the Congolese Maï Maï and Interhamwe militias, who were not signatories to peace agreements in 2001, seized control of forest areas, logged timber and looted parks and forest reserves (Hart & Mwinyihali, 2001).

The lucrative trade in plundered timber may make it financially attractive for all sides of a conflict to perpetuate the war economy and postpone demobilization. Such a scenario emerged in Cambodia during the 1990s when opposing forces preferred a lucrative stalemate in which rich forests remained under their respective domains and logging continued seemingly unimpeded by outright violence (Le Billon, 2002). The UN panel of experts on the DRC similarly observed that following the peace agreements a "sporadic, low intensity conflict" dragged on in the context of a "certain status quo" in which precious resources were "extracted, traded and routed for export" (UNSC, 2001b, par. 58). If conflict timber continues to fund insurgents, state militaries and militias in the DRC and Liberia, long-lasting peace, disarmament and sustainable forest management could remain elusive.

4 ADDRESSING CONFLICT TIMBER

A spectrum of mechanisms and schemes are available to address the conflict timber problem, including international treaties, conventions, certification regimes, monitoring technologies, bilateral and regional agreements, UN sanctions, and the influence of conditions tied to development assistance funds. The different political, commercial, governance, and logistical contexts in which conflict timber has been produced and traded indicates that no one solution or combination of solutions will be appropriate for all cases. Rather a case-by-case approach should be applied at the same time that broad international capacities and options are developed. This section considers some of these options and evaluates their potential utility.

4.1 International Regulations, Treaties and Conventions

At present, no international law specifically prohibits the trade of conflict timber. Neither is there an institution or authority that is charged with regulating the trade in conflict commodities.[9] Numerous international bodies have initiatives for monitoring illicit commodities, but none possess broad powers to track the trade of natural resources or the flow of the proceeds they generate worldwide (Winer & Roule, 2002). While some of the efforts address trade and governance issues that are common to forest crime and conflict timber, most have not taken into consideration the particular political, logistical, and financial scenarios that emerge during violent conflict. For example, the Convention on International Trade in Endangered Species (CITES) and other mechanisms, assume that governments control the territory within their borders, are legitimate sovereign states and operate in the best interests of their citizens. These

[9] For detailed discussions of enforcement instruments, see Le Billon (2002) and Winer and Roule (2002).

assumptions do not hold for many states plagued by armed conflict or those where complicit government authorities frustrate the rule of law.

The approaches used to address conflict timber situations should be tailored to the circumstances of each case. However, the development and application of adequate national and international controls over financial transactions in public and private financial institutions is necessary for increasing scrutiny and control of the movement of profits from conflict timber. Harnessing the investigative and prosecutorial powers of the International Criminal Court (ICC) cases could prove useful in some extreme cases. Established in 2002 to investigate war crimes and genocide, the ICC has already recognized the link between the natural resources and violent conflict. In 2003, it announced plans to investigate companies suspected of dealing in diamonds originating in the Democratic Republic of the Congo that have been used to finance war (BBC, 2003). Falling into the same category, international traders in conflict timber could also be subject to investigation.

4.2 Certification, Timber Tracking and Verification

Currently the international community lacks a system to identify or track conflict timber or illegally logged wood as it moves through the long commodity chain from forest to consumer. This is an important consideration given that when conflict timber enters the international market it is "de facto immediately laundered in the legal timber trade" (Global Witness, 2002b, p. 44). The development of an international certification regime to control the flow of "conflict diamonds", has helped elevate one conflict commodity onto the government agendas worldwide, and led to calls for a similar system to regulate conflict timber. Unlike easily smuggled diamonds, timber products and the logging operations are much more conspicuous and thus could be more easily monitored remotely or on the ground.

One potential approach to controlling would be mandatory international form of certification known as "legal verification" or "validation". According to such a system, all timber trade between participating countries would require an independently verified certificate guaranteeing that the timber was of legal origin (cf. Rochefordiere & Mitchell, 2001). To be effective a verification or validation should also avoid the apparent principal weaknesses of the certificate regime for diamonds, namely, the reliance on voluntary participation and the lack of an independent authority to monitor and enforce rules. Nonetheless, the current concern for conflict diamonds and the implementation of a certification regime should enhance recognition of the need for a collective approach to the problem of conflict timber (Renner, 2002; Winer & Roule, 2002).

Developing effective mechanisms for chain of custody tracking are essential first steps to creating a working verification system. Participating governments and monitoring groups would need to be able to track timber from the forest to the overseas market (Global Witness, 2002b). Experience from forest certification schemes, such as the Forest Stewardship Council, indicates that implementing a cost-effective chain of custody faces significant challenges, even in peaceful, fairly stable market environments.

Given the severe economic and governance problems in both conflict and postconflict scenarios, the development and implementation of a verification system would require substantial and sustained technical and financial support from industrialized nations and the donor community. The efficacy of a verification system would depend on how strictly countries enforced the rules. Voluntary bilateral accords between producer and consumer countries on the verification of timber origin and chain of custody tracking could form the groundwork for more comprehensive multilateral international agreements on chain of custody tracking and verification (Global Witness, 2002b).

In contrast to the legal focus of verification schemes, "forest certification" systems involve the voluntary participation of producers who, in demonstrating relatively good forest stewardship, seek to improve market share and the long-term viability of their forestry enterprise. For those companies specializing in working amidst political instability, where corruption and poor governance are the norm, voluntary certification is likely of no interest. As stated above, the armed groups and logging companies involved in the conflict timber trade have largely short-term goals, namely the rapid accumulation of capital and the maintenance of political power.

4.3 Regional Initiatives

A familiarity with local political and economic forces in the region as well as proximity to the problem would seem to give regional organizations an advantage in addressing issues such as conflict timber. Within the African context, however, the effectiveness of regional organizations in addressing this problem has been limited. The experience of the Economic Community of West African States (ECOWAS) in Liberia, highlights the potential problems. In 1993, the military arm of ECOWAS, known as the ECOWAS Monitoring Group (ECOMOG), responding to the UN Security Council's failure to expand the arms embargo, imposed a military blockade of one of Liberia's main ports from which the National Patriotic Front of Liberia (NPFL) exported many resources, including timber. However, ECOWAS actions failed to stop the traffic of diamonds and timber due to poor regulatory and enforcement capacity and political divisions among member states, the latter reflecting in large part commercial interest in the trade (Le Billon, 2002).

In Asia, where Myanmar might be considered an illustration of the failure of regional concern, the Association of Southeast Asian Nations (ASEAN) professes a policy of "constructive engagement". Conflict timber, however, is an issue yet to be broached at all seriously with Myanmar, a relatively new member of ASEAN. Whereas interethnic tensions between the dominant Burman people and numerous minority groups in Myanmar may have sparked various separatist rebellions, conflict timber has to a great extent fuelled this long simmering conflict. Groups along Myanmar's borders have been active in trading not only timber but wildlife species and illicit drugs to fund their operations (Marshall & Davis, 2002).

To date, the role of logging in the political economy of conflict, peace and transition in Myanmar has received scant international attention. The role that more stable neighboring countries have played in perpetuating conflict in Myanmar and profiting from its political instability should be the subject of international concern. Cease-fire deals, often associated with timber extraction, have failed to address the underlying ethnic conflicts that relate to marginalization and discrimination against minority groups in this country. The ragged end to the conflict has left Myanmar's forests exposed to ongoing exploitation by remnants of armed militias and government forces. This experience should provoke concern for the use of forest resources in the negotiation of cease-fires and peace settlements and provide warning of the danger when valuable forest resources are ceded to groups permitted to remain armed and unaccountable to civilian control.

4.4 The United Nations Experience

Leadership from the UN has helped to raise the profile of conflict timber and chart an agenda for strong and concerted international action. In several reports released in 2001, the UN Secretary General called upon the Security Council and UN member states to consider targeted measures against parties to armed conflict, including neighboring countries, whose actions contribute to the illegal exploitation and trade of natural resources that fuel of violent conflict (IPA & Fafo, 2002). With the ban imposed on Liberian logs and timber products in 2003, member states and the UN demonstrated an increasing commitment to confronting the conflict timber trade in Africa.

The design and justification of sanctions regimes for the DRC and Liberia have been possible due to the work of the UN Security Council's expert panels. They have provided some of the most credible and thorough reports on the methods and actors involved in the conflict commodities trade in sub-Saharan Africa. Although the panels are temporary bodies and work overtly and with limited mandates, they have "significantly affected the operations of sanctions busters, by indirectly curtailing access to financial credit, forcing them to change their logistical base or to seek protection in friendly countries" (Le Billon, 2002, p. 23). The achievements and limitations of the panels suggests that more progress on the worldwide problem of conflict commodities could be made by establishing a permanent and more robustly funded UN office charged with aggressively investigating cases of conflict commodities and the illicit resource exploitation in the context of armed conflict.

Although UN sanctions are an important tool for addressing the trade of conflict commodities, they are not appropriate in all situations. Their effectiveness can be eroded by systematic violation by smugglers, ineffective enforcement and political opposition from commercial interests and governments (Le Billon, 2002). Vested interests may explain why a UN ban on Liberian conflict timber was postponed for several years despite mounting evidence of timber industry involvement in the illicit arms trade (Global Witness, 2002c; UNSC, 2001a, 2001b, 2001c). Despite this mixed record and predictable opposition, sanctions remain one of the only truly forceful international tools for rapid restraint of the trade in conflict commodities.

Sanctions, which may take the form of partial or total bans, can restrict the trade of conflict commodities and reduce the funds available to armed groups and militaries.

Proposed sanctions, such as a ban of log exports from Liberia, have been criticized because of their potentially negative effects on employment and legitimate industry in a country already facing severe economic circumstances. Proponents of temporary restrictions on timber exports have dismissed concerns over their potential social and economic impacts as exaggerated, arguing that the humanitarian costs of allowing the trade of conflict timber to continue far surpass the economic costs associated with a temporary ban on log exports (Global Witness, 2002c; Smillie, 2002).

One alternative to a blanket ban on timber exports is the so-called "smart sanctions" option that focuses on restricting the activities of specific individuals and corporations most closely involved in conflict timber rather than entire industries or product lines. The UN panel of experts on the DRC has recommended that such targeted sanctions be placed on numerous companies and individuals that it has identified as party to illicit resource extraction and trade (UNSC, 2001b, 2001c). These sanctions make strategic and political sense where conflict timber represents only a small fraction of forestry sector operations and broader bans would suppress legitimate forestry activities and create strong political opposition.

4.5 Other International Initiatives

International donor agencies and lending institutions like the World Bank and the International Monetary Fund (IMF) have the potential to exercise considerable influence on client countries that lack sufficient political resolve or economic motivation to address the involvement of their companies, militaries and citizens in the conflict timber trade. Conditional disbursement of aid may elicit cooperation from states that produce and export conflict timber or are complicit in or tolerant of its trade, transport, processing, or financing. Although conditions may be applied at any time, they may be most effective in the context of peace negotiations and postconflict reconstruction, when international support is needed to ease the transition of governments and insurgents into postwar economic and political roles. Past experience indicates that the effectiveness of conditions as means of controlling conflict timber depends upon several factors, including the importance of aid to the recipient states and the cooperation and cohesion of institutional lenders (Le Billon, 2002).

One of the main factors that hastened the end of the conflict timber trade in Cambodia was Thailand's closure of its border to Cambodian logs. The role that international pressure played in eliciting Thailand's cooperation provides lessons for addressing the cross-border trade of conflict timber in Africa and elsewhere. Without the international threat of severe financial consequences for Thailand, it is doubtful that the trade would have been curtailed. The efficacy of international pressure on Thailand contrasts with the results of the IMF's attempt to influence Cambodia's conflict timber trade following the widely publicized timber deal between the Royal Cambodian Government and Thailand. In 1996, the IMF put its loan program on hold, and in the following year, canceled its Cambodia program (Global Witness, 2002b; Le Billon, 2000). While this bold decision raised the international profile of forest governance and conflict timber issues, it produced few tangible results as both the military and the Khmer Rouge continue to log forests under their control (Le Billon, 2000). On the other hand, donor pressure on Thailand was important in closing the Cambodian-Thai border thus throttling the flow of conflict timber from Cambodia.

5 CONCLUSIONS

Efforts to control the extraction and international trade of conflict timber face many logistical and political challenges both on the ground and within the international political arena. Economic interests tied to the processing, transport, sale and consumption of conflict timber make it politically difficult for many states—including those within and far from regions affected—to support action to curtail the trade. The political challenge is intensified by low level of awareness of the conflict timber within many policy-making bodies and the popular media. Even when political support exists, the remoteness of most extraction areas and the clandestine nature of the trade make systematic intelligence gathering, market surveillance and enforcement particularly difficult. Given such conditions and various levels of complicity of militaries and state authorities, neighboring governments, timber companies and rebel groups, it is necessary to tailor policy responses to individual cases.

Where the military and business interests are essential to maintaining political power, government leaders will likely resist efforts that could result in restricting the benefits that these groups receive from the trade in conflict timber. Without willing cooperation from state authorities or the armed factions that control forested areas, it may be very difficult to carry out effective monitoring of the logging activities that supply conflict timber. Weak government resolve in countries that produce and export conflict timber is compounded by problems relating to poor governance, limited capacity for monitoring and enforcement, and inadequate regulations.

The difficulties in monitoring and controlling conflict timber in producer and neighboring transit countries heighten the responsibility and potential role of consumer countries in addressing this problem. With greater economic and technical resources as well as political stability, importing countries can support monitoring and enforcement, promote multilateral mechanisms for regulating trade and finance, and promulgate domestic policies on the importation of conflict timber and dealings with companies associated with conflict commodities.

Addressing conflict timber will require both short- and long-term approaches. Ongoing and rapidly evolving conflict situations require ready and flexible international responses that will have some immediate effect on the trade and finance of conflict timber. Developing and implementing an international framework for regulating conflict timber will require concerted and sustained commitment on the part of the international community. These mechanisms remain only stopgap measures, however, unless the underlying factors encouraging conflict and misuse of natural resources, namely, poor governance, inequitable resource distribution and inadequate forest regulation, are also addressed. Preventing the emergence of new conflict timber scenarios should be of key concern for governments and international organizations addressing current crises. The established links between improved governance and community-based sustainable forest management suggest the potential that exists for forestry initiatives to simultaneously address underlying factors of conflict timber and promote broader goals of development and democracy. As a final note it should be emphasized that all experience indicates that there is no "quick-fix", no panacea for the problem of conflict timber. Each situation is different, and each requires careful analysis to determine the particular local, national, regional and international mechanisms at work and that assist in perpetuating the conflict. An understanding of the interplay of the complex variety of factors—history, ethnicity, economic, among others—will be crucial to tailoring specific policies and programs needed to address successfully the problem of conflict timber.

REFERENCES

BBC. (2003). *Firms Face "Blood Diamond" probe*. Sept 23, 2003. http://news.bbc.co.uk/go/pr/fr/2/hi/business/3133108.stm.

Collier, P., Hoeffler, A., & Söderbom, M. (2001). *On the Duration of Civil War*. Policy Research Working Paper No. 2681. Washington, D.C.: The World Bank.

Forests Monitor. (2001). *Sold Down the River: The Need to Control Transnational Forestry Corporations: A European study.* Cambridge, UK: Forests Monitor, Ltd.

Global Witness. (2002a). *Branching Out: Zimbabwe's Resource Colonialism in Democratic Republic of Congo*, 2nd ed. London: Global Witness.

Global Witness. (2002b). *Logs of War: The Timber Trade and Armed Conflict.* Fafo Report No. 379. Oslo: Fafo Institute for Applied Social Science, Programme for International Cooperation and Conflict Resolution.

Global Witness. (2002c). *Logging Off: How the Liberian Timber Industry Fuels Liberia's Humanitarian Disaster and Threatens Sierra Leone.* London: Global Witness.

Hart, T., & Mwinyihali, R. 2001. *Armed Conflict and Biodiversity in Sub-Saharan Africa: The Case of the Democratic Republic of Congo (DRC).* Washington, D.C.: Biodiversity Support Program.

International Peace Academy and the Fafo Institute for Applied Social Science. (2002). *Economic Agendas in Armed Conflict: Defining and Developing the Role of the UN.* Papers presented at the symposium at the UN Millennium Hotel, New York, NY, Mar 25, 2002.

Le Billon, P. (2000). The political ecology of transition in Cambodia 1989–1999: War, peace and forest exploitation. *Development and Change, 31*(4), 785–805.

Le Billon, P. (2002). International Instruments of Enforcement. Paper presented at *Conference on The Governance of Natural Resources Revenues*, Paris, France, December 9–10, 2002.

Marshall, A., & Davis, A. (2002). Soldiers of fortune. *TIME*, December 16, 2002, pp. 16–24.

McNeely, J. A. (2003). Biodiversity, war and tropical forests. In S. V. Price (Ed.), *War and Tropical Forests: Conservation in Areas of Armed Conflict* (pp. 1–20). New York: Haworth Press, Inc.

Renner, M. (2002). *The Anatomy of Resource Wars*. Worldwatch Paper No. 162. Washington, D.C.: Worldwatch.

Rochefordiere, A., de la, & Mitchell, A. (2001). *Independent Validation of Legal Timber: First Steps Towards a Legal and Sustainable Timber Trade.* Geneva: SGS.

SAMFU. (2002). *Plunder: The Silent Destruction of Liberia's Rainforest.* Monrovia, Liberia: The SAMFU Foundation.

Sato, H., Yasui, K., & Byamana, K. (2000). Follow-up survey of environmental impacts of the Rwandan refugees on eastern DR Congo. *Ambio, 29*(2), 122–123.

Smillie, I. (2002). *Diamonds, Timber and West African Wars*. Ottawa: International Institute for Sustainable Development (IISD).

Thomson, J., & Kanaan, R. (2004). *Conflict Timber: Dimensions of the Problem in Asia and Africa. Vol. 1. Synthesis Report.* Burlington: ARD, Inc.

United Nations Security Council (UNSC). (2001a). *Report of the Panel of Experts Pursuant to Security Council Resolution 1343 (2001),* Paragraph 19, Concerning Liberia, S/2001/1015, October 26, 2001.

United Nations Security Council (UNSC). (2001b). *Report of the Panel of Experts on the Illegal Exploitation of Natural Resources and Other Forms of Wealth of the Democratic Republic of the Congo*, S/2001/357, April 12, 2001.

United Nations Security Council (UNSC). (2001c). *Addendum to the Report of the Panel of Experts on the Illegal Exploitation of Natural Resources and Other Forms of Wealth of the Democratic Republic of the Congo*, S/2001/1072, November 13, 2001.

Wilkie, D., Shaw, E., Rothberg, F., Morelli, G., & Auzel, P. (2000). Roads, development and conservation in the Congo basin. *Conservation Biology, 14*(6), 1614–22.

Winer, J. M., & Roule, T. J. (2002). The Finance of Illicit Resource Extraction. Paper presented at the *Conference on the Governance of Natural Resources Revenues*, Paris, France, Dec 9–10, 2002.

CHAPTER 8

ENVIRONMENTAL DAMAGE FROM ILLICIT DRUG CROPS IN COLOMBIA

MARÍA D. ÁLVAREZ

1 INTRODUCTION

The natural habitats of Colombia, its forests, rivers, and grasslands, are global conservation priorities because of the richness and endemism of their fauna and flora (McNeely *et al.*, 1990; Myers *et al.*, 2000; Olson & Dinerstein, 1998; Stattersfield *et al.*, 1998). The ecosystem function of these natural habitats benefits millions of Colombians who depend on their water, wood, bushmeat, and medicinal plants (Rodríguez & Ponce, 1999). Over several centuries, the advance of the agricultural frontier has fragmented these ecosystems resulting in the extirpation of endemic species, natural disasters, and a general decline in environmental quality, particularly in the Andes and the Caribbean region (Cavelier & Etter, 1995; Cavelier *et al.*, 1998; Etter & van Wyngarden, 2000). Over the last decade, however, incentives for agricultural production in Colombia have decreased. The output of annual crops fell at an average annual rate of 3.2%, agricultural production lost 4% of its relative importance in the GDP, and agricultural imports have increased almost 10-fold (Jaramillo, 2001; Robledo, 1999; Vásquez, 1997). Nonetheless, the fragmentation of natural ecosystems persists partly because of the expansion of other legal economic activities—perennial crops, cattle ranching, mining, and timber exploitation—as well as that of illicit crops whose exponential growth has paralleled the escalation of armed conflict (Álvarez, 2001, 2003).

The most salient characteristic of Colombia's recent history is the generalization of armed conflict by belligerent groups—leftist guerrillas and right-wing paramilitaries—between each other and against the state and civil society. Armed conflict and the resulting sociopolitical decomposition have, in one decade, killed more than 250,000 Colombians (Comisión Colombiana de Juristas, 2000) and displaced around 1.35 million (Zuluaga, 1999). The protracted unfolding of the Colombian conflict, over almost four decades, has also brought about favorable conditions for the expansion of illicit crops and trade in prohibited drugs. To quote a World Bank report: "Government and civil society alike recognize that violence is *the* key development constraint" (World Bank, 1999, I, italics in the original).

W. de Jong, et al. (eds.), Extreme Conflict and Tropical Forests, 133–147.

At the same time trade in prohibited drugs and, more recently, the production of illicit crops have provided money, as well as political and economic influence, to fuel the conflict (Bejarano & Pizarro, 2003; Reyes 1990, 1999; Thoumi, 1995). Drug trafficking is estimated to have provided 42% of guerrilla income between 1991 and 1995 (Echandía, 1999), and is known to finance even more of the paramilitaries' income (Cubides, 1999; Reyes, 1999). Armed conflict in Colombia is not only a contest over territory in large forested areas, but of competing land management policies therein (Álvarez, 2003; Ramírez, 1998). Thus the conflict itself constitutes an independent variable directly or indirectly affecting natural ecosystems.

Land use in Colombia is directly affected by the forced displacement of local communities and environmental authorities from many regions, as well as the decrease in the rate of frontier colonization resulting from armed conflict (Álvarez, 2003; Dávalos, 2001; Ramírez, 1998). Abandoned habitats may regenerate into fallow, or be consolidated into larger pasture by large landholders, agricultural productivity decreases, and enforcement of environmental protection in protected areas becomes increasingly difficult, if not impossible (Álvarez, 2001, 2003; Andrade, 2004; Reyes, 1999). The magnitude of the forest resources in conflict is significant. Of Colombia's remaining forests, 33% are in municipalities with medium to high activity by armed groups, and 20% of them are in municipalities where both guerrillas and paramilitaries are present (Álvarez, 2003). Armed conflict also has indirect ramifications. Frontiers are virtually lawless, illicit crops become the only viable economic alternative in many regions in conflict, and land consolidation has reversed the gains of the twentieth century agrarian reform (Álvarez, 2003; Reyes, 1999). Understandably, there are virtually no incentives for taking the longer-term view in land management in the forested frontiers.

In a previous publication I outlined the obstacles to conservation arising from the belligerence of armed groups (Álvarez, 2003). This chapter focuses on the expansion of illicit cash crops, which is mostly taking place in forested areas where armed groups wield their power (Álvarez, 2001, 2002, Andrade, 2004; Reyes, 1990, 1999; Ramírez, 1998). The chapter answers questions like: What are the environmental effects of the illegal activities associated with illicit crops and the production of prohibited drugs? How important are these effects to the conservation of natural ecosystems and the economic development of the country? In doing so I summarize current knowledge about deforestation, pollution, and injuries to biological diversity arising from elicit activities. The goal of the chapter is to both stimulate research on the political ecology of armed conflict and elicit crops, and to advance toward the mitigation of these damages. Besides suggesting new directions in research, this paper aims to bring environmental variables into the mainstream academic study of the conflict and the political discussions surrounding its resolution.

2 DATA ON DAMAGES

The figures on the extension and location of illicit crops, yield per hectare, and production of coca leaf and opium in Colombia since 1986 are available from the UN-ODCCP (1999, 2000, 2001), UN-ODC (2003), and UNDCP (2000) documents. The estimates

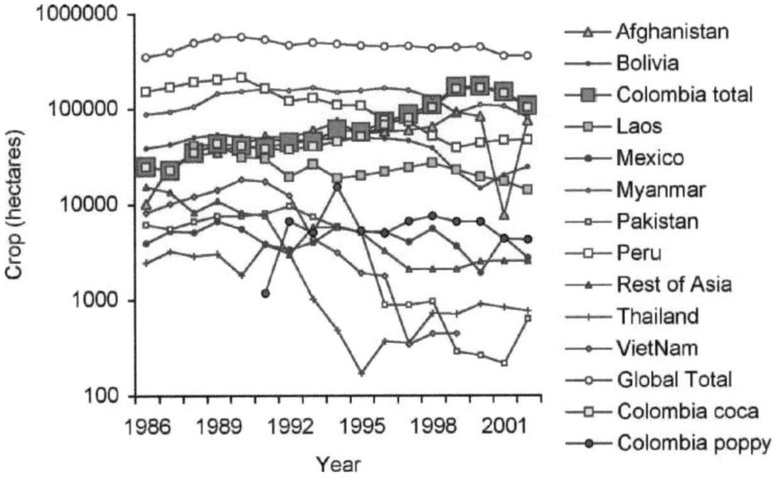

Figure 8.1. *Illicit crops in Colombia and the world since 1986 (UN-ODCCP 2001). The vertical scale is logarithmic, so the increase in illicit crops in Colombia over the study period is of one order of magnitude.*

of the quantities of reagents used in processing prohibited drugs are based on the estimates of 0.315 l gasoline and 0.002 l sulfuric acid per kg of coca leaf to produce coca base (PLANTE, 1996). The average yield is 0.0088 kg coca base per kg of coca leaf. One kilogram of coca base requires the use of 56.775 l of acetone or ethyl ether (Elsohly *et al.*, 1984; Schlesinger, 1985). The reagents for opium processing were estimated at 0.5 l alcohol, 0.5 l ethyl ether, and 5 l hydrochloric or sulfuric acid per kg of opium processed into heroin (Narayanaswami, 1985). The geographic analyses on the distribution of forests (Álvarez, 2003) and richness of threatened and endemic birds of Colombia (Álvarez, 2002) presented here are published in detail elsewhere.

2.1 Deforestation

Figure 8.1 shows the extension of illicit crops in Colombia and all producing countries since 1986. According to Cavelier and Etter (1995), 85% of poppy crops are planted in newly deforested lands and additional area of 2.5 to 3.0 times the cropped area is cleared for food crops and airstrips. According to the narcotics police each hectare of coca planted implies the clearing of four hectares of forest (El Tiempo, 2001; Nyholm, 1998). Beyond the deforestation from illicit crops per se, there are additional damages associated with chemical eradication of these crops in the range of several tens of thousand hectares (Figure 8.2).

2.2 Liquid Effluents

It is difficult to estimate with precision the amount of liquid waste resulting from the processing of narcotics. Conditions vary from one laboratory to another making anecdotal

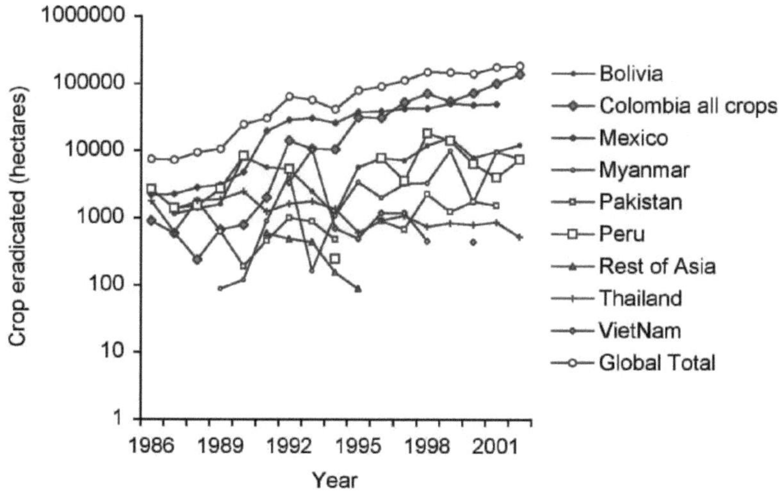

Figure 8.2. *Eradication using chemical herbicide in Colombia, by various methods and in other countries. The vertical scale is logarithmic, so the increase in illicit crop eradication using glyphosate in Colombia over the study period is of more than one order of magnitude.*

figures (e.g., Armstead, 1992; Dávalos, 2001) useless in generalizing to an estimation of the liquid waste for overall production. Additionally, the use of volatile substances complicates the calculation of effluents. Using the estimates described in the section above, the use of solid and liquid chemical reagents is shown in Table 8.1. This table overestimates the amount of waste generated in coca leaf to coca base processing for the year 2000. The processing of coca base to cocaine, which uses hydrochloric or sulfuric acid and ammonia, has not been included for lack of reliable published estimates.

2.3 Biological Diversity

Although numerous publications on illicit crops refer to biodiversity (e.g., USDS, 2001), except for Cavelier and Etter (1995), no analyses until those of Álvarez (2002, 2003) had presented ecosystems or species affected by illicit crops. Figure 8.3 shows

Table 8.1. *Estimation of the amounts of solid and liquid chemical reagents used in processing illicit crops during the year 2000.*

Coca leaf (Tonnes)	*Cement (Tonnes)*	*Sulfuric acid (liters)*	*Gasoline (liters)*	*Coca base (Tonnes)*	*Acetone (liters)*	*Total liquids (liters)*
40,354	33,494	80,709	12,711,636	355	20,161,865	32,954,210
Opium (Tonnes)	Alcohol (liters)	Ether (liters)	Acid (liters)			Total liquids (liters)
88	44,000	44,000	440,000			528,000

Sources: Elsohly *et al.* (1984), Narayanaswami (1985), PLANTE (1996), Schlesinger (1985), UNDCP (2000), and UN-ODCCP (1999, 2000, and 2001).

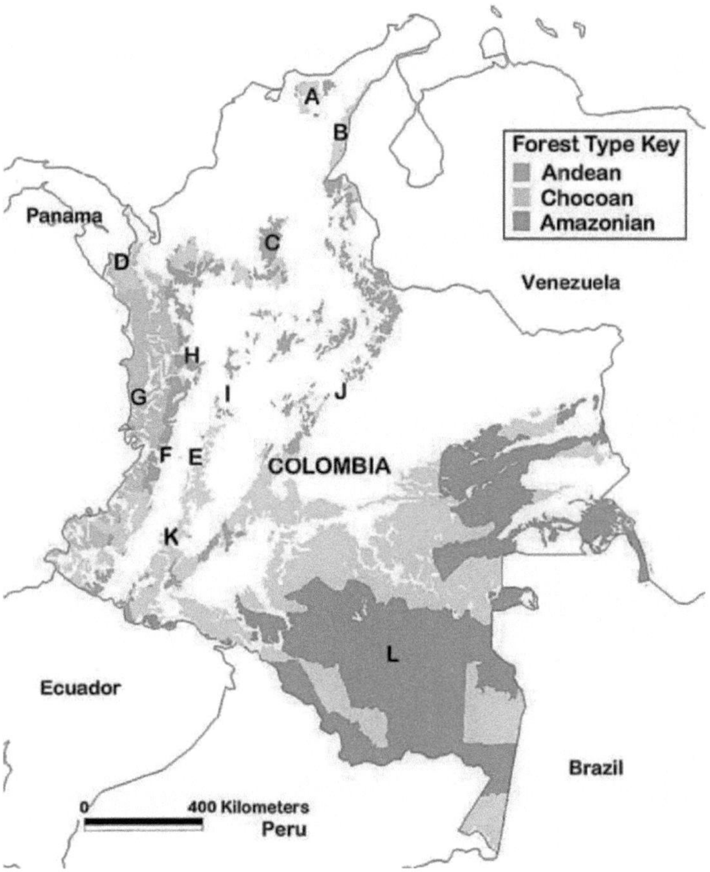

Figure 8.3. *Forests in municipalities where illicit crops have been found. Main regions mentioned in text: A, Sierra Nevada de Santa Marta; B, Serranía del Perijá and Serranía de los Motilones; C, Serranía de San Lucas (Central Andes); D, Darién lowlands; E, Páramo de Las Hermosas (Central Andes); F, Farallones de Cali (inter-Andean slope of West Andes); G, Chocoan forests; Andes: H, West; I, Central; J, East; K, Macizo Colombiano, and L, Amazonia.*

three types of forests, Andean, Chocoan and Amazonian, in municipalities where illicit crops have been reported (Reyes, 1999). Figure 8.4 shows the number of threatened bird species in relation to forests in municipalities where illicit crops have been reported. These results do not account for the damages to aquatic ecosystems caused by chemical waste described in the previous section.

3 HOW MUCH DEFORESTATION CAN BE BLAMED ON ILLICIT CROPS?

The growth of illicit crops (Figure 8.1) and the expansion of eradication programs using chemical defoliants (Figure 8.2)—which also cause instant defoliation in forested areas adjacent to illicit crops—are alarming. The situation is even more pressing if the

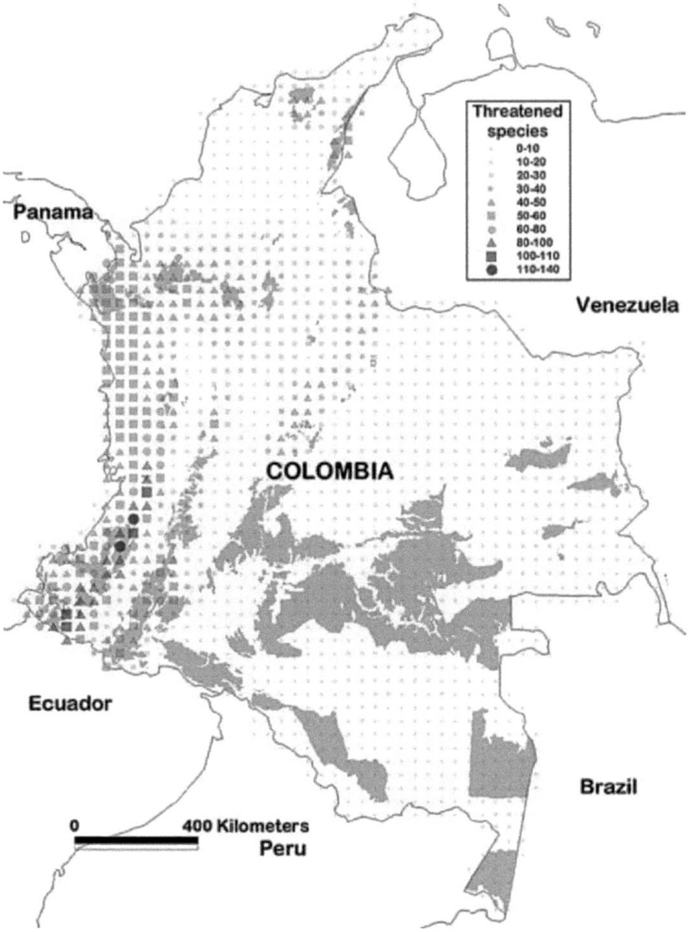

Figure 8.4. Number of threatened bird species according to Brooks et al. (1999) and forests in municipalities where illicit crops have been found (gray).

deforestation associated with illicit crops (2.5 to 4 times the area planted, see previous section) is considered. Nevertheless, both the figures on illicit crop production and associated deforestation have important margins of error worth discussing.

In particular, the multiplication factors used in estimating deforestation caused by illicit crops have not been studied critically. The estimation of Cavelier and Etter (1995) dates back to 1992 and the factors associated to coca cited by El Tiempo (2001) and Nyholm (1998) are not supported by published empirical research. To illustrate how these estimations can be misleading, one must examine the estimation of overall deforestation in Colombia. Between 1990 and 1995 Colombia lost 1,311,000 hectares of forest according to the FAO (1999). If the factors relating deforestation to illicit crops are correct then illicit crops caused 70.4% of the deforestation over that period.

That is to say, the expansion of legal crops and cattle ranching into forest frontiers, mining and timber exploitation and other miscellaneous uses caused only 29.6% of the deforestation of the country. If we take the average annual deforestation rate of 190,000 ha for the decade of the 1990s (FAO, 2005), then illicit crops caused 1.7 times the deforestation reported. This is impossible. Either the overall deforestation figures are incorrect, and given the limitations of such global assessments this is likely, or the estimation of multiplication factors relating deforestation to illicit crops is flawed, or some combination of both.

Illicit crops are one of the main, perhaps the main, cause of deforestation in Colombia. Nevertheless, it is impossible to establish how important they are without knowledge of deforestation caused by legal activities. Besides basic surveys measuring local and regional forest cover, research on the dominant means of production is necessary to understand the contribution of illicit crops and armed conflict to general deforestation. A historical view of the evolution of illicit crops is indispensable to anticipate the expansion of this threat.

As with any other agricultural commodity, soil quality, land availability and transportation determine the locations and methods used to cultivate these crops. By replacing less profitable agricultural activities, illicit crops may produce less deforestation than their legal counterparts (Henkel, 1995; Kaimowitz, 1997; Young, 1996). At the same time, illicit crops may be profitable enough to bring into production lands that would not be exploited otherwise, and attract migrants to agriculture who were engaged in other economic activities, thus increasing deforestation (V. Tafur, personal communication) (Table 8.2).

Table 8.2. *Area affected by illicit crop production and processing, 1998.*

Type of ecosystem	*Locality*	*Legally protected areas and indigenous reserves*	*Estimated damage (ha)*
Tropical humid forest			75,300
	Amazonian foothills	NNP Picachos*	66,800
	Serranía de San Lucas	No protected areas	6,500
	Magdalena Medio	No protected areas	2,000
Andean forests			16,127
	Sierra Nevada de Sta. Marta	NNP and IR Sierra Nevada	750
	Serranía de Perijá	NNP and IR Catatumbo–Barí	7,800
	Serranía de los Motilones	IR Motilonia	950
	West Andes	NNP Farallones	3,827
	Nevado del Huila	NNP Nevado del Huila	300
	Macizo Colombiano	NNP Puracé and Munchique	2,500
Gallery forests			12,000
	Apaporis river galleries	IR Nukak, Vaupés River	10,000
	Other rivers	NNP La Paya	2,000
Total			108,077

Sources: Etter (1998) and UNDCP (2000).

*NNP, national natural park (government protected area); IR, indigenous reserve (community managed area).

Tentatively, the rich high Andean soils near densely populated areas where opium poppy is being grown in the *Macizo Colombiano*, and the *Páramo de las Hermosas* would undergo further forest clearing if other crops—e.g., potatoes or wheat—were used. Conversely, opium poppy growing in the *Serranía del Perijá* conceivably could not be replaced by any other licit crop, given the distances to markets. This assessment assumes that armed groups exerting authority in these areas would be willing to allow the switch to less profitable crops, and that market pressure for legal replacement crops would supersede environmental law enforcement in the many protected areas affected by poppy in these forested regions (Table 8.2). It is even more difficult to speculate about the lowland regions where coca currently grows. Soil quality and distances to market would make most agriculture unprofitable. But migrants to these regions have already endured numerous boom-and-bust cycles of which coca production is only the latest and bloodiest (Ramírez & Molano, 1998). Therefore trying to assess their environmental choices based on purely economic reasons rather than, e.g., on the political economy of land tenure (Reyes, 1990, 1999), is a futile exercise. In the absence of empirical research, these statements inevitably pose more questions than they answer regarding illicit crops and armed conflict.

4 WHY ARE STUDIES ON LEGAL AND ILLICIT WASTE MANAGEMENT NECESSARY?

The environmental effects of the chemical waste generated by the processing of prohibited drugs have not been studied with empirical data. The figures presented here (Table 8.1), as those shown in previous exercises (Armstead, 1992; Dourojeanni, 1992; Osorio-Bryson, 1992; USDS, 2001), are calculated based on production or purchases of chemicals and cannot replace field studies of actual effluvia. Rather, they are meant to draw attention to this problem. Precisely to call attention Armstead (1992), Dourojeanni (1992), and USDS (2001) ascribe a disproportionate environmental impact to illicit crops without presenting evidence to show these effects to be larger, smaller, or equal to those generated by legal activities, including illicit crop eradication. How many of the landslides, floods, and other natural disasters mentioned in Armstead (1992), Dourojeanni (1992) or USDS (2001) can be attributed to illicit crops? This deserves noting because deforestation (Cavelier *et al.*, 1998; Etter & van Wyngarden, 2000), soil erosion (Sarmiento, 2000), and the pollution of watercourses are neither recent nor are they only brought about by illicit production (for an example of background rates of water contamination see Squillace *et al.*, 1999).

The gaps in our knowledge on the processing and management of wastes generated by the production of prohibited drugs are enormous. We ignore: (1) What regional adaptations exist in processing systems, (2) what fraction of soluble and volatile reagents in fact reaches watercourses, (3) what is the persistence of pollutants in the water and soil, and (4) what direct and indirect effects these wastes have on the fauna and flora. Preliminary results using fuels show accumulation of some organic compounds in groundwater (Pasteris *et al.*, 2002), but the rates of biotic degradation are probably highly site-specific (Barber *et al.*, 2001). More importantly, studies measuring impact on the local aquatic fauna and flora are available only at the most general level (i.e., see Vargas, 1999; Velaidez, 2000). We are also ignorant about the context in which these wastes are placed. Dumping 50 l of chemicals onto the lower Putumayo, the river that forms the Colombia – Peru border, is not likely to have the same affect as discarding them on a mountain stream. Moreover, how important are these wastes compared to legal pollution in the producing localities, in other regions, or in the country? The estimates shown in Table 8.1 are a necessary but insufficient step toward answering these questions.

5 EFFECTS ON BIOLOGICAL DIVERSITY

All the forest types considered here (following Etter, 1998) located in the Andes, Chocó, and Amazon have been affected by illicit crops. Most of these crops, however, are located in Amazonian forests (Figure 8.3, Table 8.2). The risks of species loss arising from deforestation, mainly from poppy cultivation, however, are more imminent in the Andean region where only 27% of the original forest cover remains (Cavelier & Etter, 1995). This risk became particularly acute with the convulsions in the world market for heroin following the Taliban-led eradication of opium poppy in Afghanistan (Crossette, 2001), and the subsequent course of events there. In 2006 poppy production in Afghanistan remains high for its third consecutive year while Colombian production is half what it was in 2000–2001. The Andean forests of Colombia have been highlighted as the richest in endemic plants and vertebrates (Myers *et al.*, 2000) and all studies of global biological diversity list them as highly threatened and requiring of immediate protection (Olson & Dinerstein, 1998; Stattersfield *et al.*, 1998). These same studies also highlight the Colombian Chocó, currently threatened by coca cultivation (Figure 8.3, Table 8.2).

The proposal for investment in sustainable development in environmentally fragile areas of Plan Colombia (Presidencia de la República 2000), however, only mentions Amazonia and not the Andean or Chocoan forests as a priority for environmental management. Without detracting relevance from the efforts to conserve Amazonian forests, the oversight of biological diversity in planning (and publicizing) the environmental strategy of Plan Colombia is frightening. Environmental decisions linked to the budget of Plan Colombia are changing the exploitation of forest resources and therefore the biological diversity of all the *departamentos* in southern Colombia (Álvarez, 2001). Ignorance or lack of interest in the ecosystem function and biodiversity of the different regions at stake can lead to a reprise of the history of the last two centuries. Human impact has completely cleared the Caribbean forests, reduced the Andean forests to their current extension, and made inroads into Amazonia and the Chocó. The decline in environmental quality and environmental disasters associated with these changes should alert all conservationists in Colombia and the world to restrain the continuation of such so-called development in this century.

The most important areas for bird conservation affected by illicit crops are the southern Andes, the northern West Andes, the lowlands of Darién, the *Sierra Nevada de Santa Marta*, the *Serranía del Perijá*, and the *Serranía de San Lucas* (Figure 8.4).

The forests with largest areas threatened by illicit crops are in Amazonia and the Amazonian foothills of the East Andes (Table 8.2), with relatively low priority for avian conservation, but important for numerous mammal species (Voss & Emmons, 1997).

The current trends in expansion and eradication of illicit crops (Figure 8.1) and the restricted distribution of many species might result in several extirpations, and possibly a few extinctions, among endemic species in affected areas. The risks of species loss are particularly acute for birds inhabiting the strip of mountainous forests threatened by opium poppy in the *Serranía del Perijá*, *Macizo Colombiano*, and the West Andes of Cauca and Nariño, as well as the Chocoan humid forests of Valle, Cauca, and Nariño. Despite the greater size of damage in Amazonia, birds there may be facing lower risk because their remaining natural habitat is relatively large (see Álvarez, 2002 for an in depth discussion on this point).

One must emphasize here that extinction risks make no distinction between legal and illegal causes of habitat fragmentation. Deforestation caused by eradication using the aerial herbicide glyphosate, by crop substitution programs that require greater planted area, or by the construction of infrastructure as part of alternative development plans, has the same effects as that caused by illicit crop production. The risk of extinction is mostly determined by the size of remaining habitat, the permanence of deforestation, and the pattern of deforestation (total versus patchy). These characteristics maximize the potential impact on biodiversity of area-extensive crops, the increase in colonization tied to roads, and licit economic activities requiring complete clearing such as cattle ranching. Secondarily, the use of chemicals for preparing the soil, controlling pests, illegal processing, and eradication itself affect susceptible plants and animals, exacerbating the impact of deforestation and contributing to extinction (American Bird Conservancy, 2001).

6 CONCLUSIONS

The expansion of illicit crops and drug trafficking enterprises cannot be viewed in isolation from the widespread violence that it causes and where it thrives (Bejarano & Pizarro, 2003). This violent conflict is one of the numerous social and political costs of an overly narrow scope of policy alternatives in managing illicit crops and drug trafficking (Thoumi, 1995). Even when studied as a few of several causes leading up to the partial collapse of the Colombian state, both drug trafficking and US antidrug policies in the Andean region have been crucial to the deterioration of state authority in Colombia (Bejarano & Pizarro, 2003, p. 18). This paper has focused on environmental damages, but the implications of the figures presented here for society as a whole are obvious: Antidrug policies pursued so far have been ineffective and urgently need revision.

Environmental damages caused by illicit crops and processing of prohibited drugs are alarming. They affect hundreds of thousands of hectares—perhaps a few million hectares over the last 20 years—through deforestation as a result of illicit crop production and eradication using glyphosate. They pollute numerous water catchments through millions of liters of chemical reagents dumped, and threaten endangered and endemic species in aquatic and terrestrial habitats. Nonetheless, the absolute and relative magnitude of these damages is unknown because of the lack of empirical research on environmental impacts in the producing localities. Without this context the rhetoric on environmental damages caused by illicit activities cannot even provide an order of magnitude to measure these environmental damages and thus justify immediate action. This is particularly important in Colombia, where society already tolerates urban and rural environmental degradation and the entities in charge of environmental control lack the political clout and budget to fully accomplish their mandates (Rodríguez, 1998).

Until now domestic and international measures against the production of prohibited drugs in Colombia have generated enormous economic incentives for trafficking (Thoumi, 1995, 1999) and exacerbated the pressure for colonization of new lands as a response to government repression (Reyes, 1999). The environmental protection goals of Colombia would be better served by the search for economic alternatives to increase profitability and employment generation in licit activities (Álvarez, 2001). Given chronic economic depression and unemployment in the agrarian sector (Robledo, 1999; Vásquez, 1997), these alternatives must come not only from rural, but urban development (Jaramillo, 2001). In any case, in order to conserve the ecosystems that provide water and other environmental benefits to Colombia, this search for economic alternatives must consider the environmental risks arising from deforestation, pollution and the loss of species produced by both licit and illicit activities.

ACKNOWLEDGMENTS

A nonexhaustive list of researchers who have helped in this pursuit includes: The man in the Wellington boots, G. Andrade, L. G. Baptiste, J. Bates, A. M. Bejarano, S. Carrizosa, D. Donovan, ADR, EDA, J. Fjedså, W. de Jong, D. Kaimowitz, D. Kwan, M. V. Llorente, C. McIlwaine, G. Martin, C. Padoch, N. Peluso, M. Pinedo, A. L. Porzecanski, S. Price, A. Reyes, M. Rodríguez, K. Redford, E. Sanderson, V. Tafur, and K. Willett. Financial support from Columbia University is gratefully acknowledged.

REFERENCES

Álvarez, M. D. (2001). Could peace be worse than war for Colombia's forests? *The Environmentalist, 21*(4), 305–315.

Álvarez, M. D. (2002). Illicit crops and bird conservation priorities in Colombia. *Conservation Biology, 16*(4), 1086–1096.

Álvarez, M. D. (2003). Forests in the time of violence: Conservation implications of the Colombian war. *Journal of Sustainable Forestry, 16*(3/4), 49–70.

American Bird Conservancy. (2001). *Position Statement on Glyphosate Spraying in Colombia for the Eradication of Illicit Crops*. Washington, D.C.: American Bird Conservancy.

Andrade, G. I. (2004). Selvas Sin Ley. Conflicto, Drogas y Globalización de la Deforestación de Colombia [Lawless forests. Conflict, drugs and globalization of deforestation in Colombia]. In M. Cárdenas & M. Rodríguez (Eds.), *Guerra, Sociedad y Medio Ambiente* (pp. 107–173). Bogotá: Foro Nacional Ambiental.

Armstead, L. (1992). Illicit narcotics cultivation and processing: The ignored environmental drama. *UN-ODCCP Bulletin on Narcotics,* (2), 9–20.

Barber, L. B., Leenheer, J. A., Noyes, T. I., & Stiles, E. A. (2001). Nature and transformation of dissolved organic matter in treatment wetlands. *Environmental Science and Technology, 35,* 4805–4816.

Bejarano, A. M., & Pizarro, E. (2005). From "restricted" to "besieged": The changing nature of the limits to democracy in Colombia. In S. Mainwaring & F. Hagopian (Eds.), *Advances and Setbacks in the Third Wave of Democratization in Latin America* (pp. 235–260). Cambridge: Cambridge University Press.

Brooks, T., Mehlman, D., Roca, R., Jahn, A., Limp, F., Smith, K., & Williams, P. (1999). *Worldmap for Birds at Risk in Latin America.* Washington, D.C.: The Nature Conservancy.

Cavelier, J., & Etter, A. (1995). Deforestation of Montane forest in Colombia as a result of illegal plantations of Opium (*Papaver somniferum*). In P. Churchill, H. Baslev, E. Forero & J. L. Luteyn (Eds.), *Biodiversity and Conservation of Neotropical Montane Forests* (pp. 541–549). Bronx: New York Botanical Garden.

Cavelier, J., Etter, A., Aide, T. M., Santos, C., Eusse, A. M., & Dupuy, J. M. (1998). The savannization of moist forests in the Sierra Nevada de Santa Marta, Colombia. *Journal of Biogeography, 25,* 901–912.

Comisión Colombiana de Juristas. (2000). *Panorama de Derechos Humanos y Derecho Humanitario en Colombia. Informe de Avence Sobre 2000* [Overview of human right and humanitarian law in Colombia. Report of advances toward 2000]. Bogotá: Comisión Colombiana de Juristas.

Crossette, B. (2001). Taliban's eradication of Poppies is convulsing Opium market. *New York Times,* June 13, 2001, A15.

Cubides, F. (1999). Los Paramilitares y su Estrategia [The paramilitaries and their strategy]. In M. V. Llorente & M. Deas (Eds.) *Reconocer la Guerra para Construir la Paz* (pp. 151–199). Bogotá: Editorial Norma.

Dávalos, L. M. (2001). The San Lucas mountain range in Colombia: How much conservation is owed to the violence? *Biodiversity and Conservation, 10,* 69–78.

Dourojeanni, M. (1992). Environmental impact of coca cultivation and cocaine production in the Amazon region of Peru. *UN-ODCCP Bulletin on Narcotics,* (2) 37–53.

Echandía, C. (1999). Expansión territorial de las Guerrillas Colombianas: Geografía, economía y violencia [Territorial expansion of Colombian guerrillas: Geography, economics and violence]. In M. V. Llorente & M. Deas (Eds.), *Reconocer la Guerra para Construir la Paz* (99–149). Bogotá: Editorial Norma.

El Tiempo. (2001). Colombia Pierde Reserva Forestal por Tala de Cocaleros: Cultivos Ilícitos Arrasan Selvas y Bosques [Colombia loses forest reserves to coca deforestation: illicit crops clear jungles and forests]. *El Tiempo* Bogotá: May 21, 2001.

Elsohly, M. A., Arafat, E. S., Jones, A. B., Engelke, B. F., Histon, J. L. & Gentner, W. A. (1984). Study of the concentration of the herbicide (2,4-dichlorophenoxy)-acetic acid in coca leaves and paste obtained from plants treated with this herbicide. *UN-ODCCP Bulletin on Narcotics,* (2), 65–77.

Etter, A. (1998). Mapa general de ecosistemas de Colombia [general ecosystem map of Colombia]. In *Informe Nacional Sobre el Estado de la Biodiversidad 1997 Colombia.* Bogotá: Instituto de Investigación de Recursos Biológicos Alejandro von Humboldt, PNUMA, Ministerio del Medio Ambiente.

Etter, A. & Wyngarden, W. van. (2000). Patterns of landscape transformation in Colombia, with emphasis in the Andean region. *Ambio, 27*(7), 432–439.

Food and Agriculture Organisation (FAO). (1999). *The State of the World's Forests 1999*. Rome.

Food and Agriculture Organisation (FAO). (2005). *The State of the World's Forests 2005*. Rome.

Henkel, R. (1995). Coca (*Erythroxylum coca*) cultivation, cocaine production, and biodiversity loss in the Chapare region of Bolivia. In P. Churchill, H. Baslev, E. Forero & J. L. Luteyn (Eds.), *Biodiversity and Conservation of Neotropical Montane Forests* (pp. 551–560). Bronx: New York Botanical Garden.

Jaramillo, C. F. (2001). Liberalization, crisis, and change: Colombian agriculture in the 1990s. *Economic Development and Cultural Change, 49*(4), 821–846.

Kaimowitz, D. (1997). Factors determining low deforestation: the Bolivian Amazon. *Ambio, 26*(8), 536–540.

McNeely, J. A., Miller, K. R., Reid, W. V., Mittermeier, R. A., & Werner, T. B. (1990). *Conserving the World's Biological Diversity*. Gland: The World Conservation Union, World Resources Institute, Conservation International, World Wildlife Fund and The World Bank.

Myers, N., Mittermeier, R. A., Mittermeier C. G., da Fonseca G. A. B., & Kent, J. (2000). Biodiversity hotspots for conservation priorities. *Nature, 403*, 853–858.

Narayanaswami, K. (1985). Parameters for determining the origin of illicit heroin samples. *UN-ODCCP Bulletin on Narcotics,* (1), 49–62.

Nyholm, K. (1998). Narcotráfico, medio Ambiente and paz [drug trafficking, environment and peace]. In C. H. Fonseca, A. González Posso & E. Falla Duarte (Eds.), *Ambiente para la Paz* (pp. 147–148). Bogotá: Ministerio del Medio Ambiente and Cormagdalena.

Olson, D. & Dinerstein, E. (1998). The global 200: A representation approach to conserving the earth's most biologically valuable ecoregions. *Conservation Biology, 12*, 502–515.

Osorio-Bryson, L. (1992). Environment and drug trafficking. *UN-ODCCP Bulletin on Narcotics,* (2), 27–29.

Pasteris, G., Werner, D., Kaufman, K., Hohener, P. (2002). Vapor phase transport and biodegradation of volatile fuel compounds in the unsaturated zone: A large scale lysimeter experiment. *Environmental Science and Technology, 36*, 30–39.

Plan Nacional de Desarrollo Alternativo (PLANTE). (1996). Porqué son un Problema los Cultivos Ilícitos? [why are illicit crops a problem?]. *El Espectador*, Bogotá: 20 de agosto 1996.

Presidencia de la República. (2000). *Plan Colombia*. Bogotá: Presidencia de la República de Colombia. http://www.colombiaemb.org/repository/Doc_Plan_Col/Plan%20Colombia%20texto.pdf.

Ramírez, C. (1998). Conflicto agrario y medio ambiente [agrarian conflict and environment]. *Revista Foro, 35*, 5–27.

Ramírez, C. & Molano, A.. (1998). Estructura agraria, conflictos armados, cultivos ilícitos y medio ambiente [agrarian structure, armed conflictc, illicit crops and environment]. In C. H. Fonseca, A. G. Posso & E. F. Duarte (Eds.), *Ambiente para la Paz* (pp. 125–146). Bogotá: Ministerio del Medio Ambiente and Cormagdalena.

Reyes, A. (1990). La violencia y la expansion territorial del narcotráfico [violence and territorial expansion of drug trafficking]. In J. G. Tokatlián & B. M. Bagley (Eds.), *Economía y Política del Narcotráfico* (117–139). Bogotá: Ediciones UniAndes.

Reyes, A. (1999). Especial: 35 años de conflicto [special: 35 years of conflict]. Lecturas Dominicales-*El Tiempo* (Bogotá), October 17, 2–11.

Robledo, J. E. (1999). Neoliberalismo y desastre agropecuario [Neoliberalism and agricultural disaster]. *Deslinde, 25*, 32–49.

Rodríguez, M. (1998). *La reforma ambiental en Colombia: Anotaciones para la historia de la gestión pública ambiental* [environmental reform in Colombia: Notes for history of public environmental policy]. Bogotá: Tercer Mundo Editores.

Rodríguez, M., & Ponce, E. (1999). Financing the green plan ('Plan Verde') in Colombia: Challenges and opportunities. Paper presented at the workshop on financing of sustainable forest management, October 11–13, 1999, Croydon, London.

Sarmiento, F. O. (2000). Breaking mountain paradigms: Ecological effects on human impacts in managed tropandean landscapes. *Ambio, 29*(7), 423–431.

Schlesinger, H. L. (1985).Topics in the chemistry of cocaine. *UN-ODCCP Bulletin on Narcotics,* (1), 63–78.

Squillace, P. J., Moran, M. J., Lapham, W. P., Price, C. V., Clawges, R. M., & Zogorski, J. S. (1999). Volatile organic compounds in untreated ambient groundwater of the United States, 1985–1995. *Environmental Science and Technology, 33*, 4176–4187.

Stattersfield, A. J., Crosby, M. J., Long, A. J., & Wege D. C. (1998). *Endemic Bird Areas of the World: Priorities for Biodiversity Conservation.* Cambridge: BirdLife International.

Thoumi, F. E. (1995). *Political Economy and Illegal Drugs in Colombia.* Boulder, CO: Lynne Rienner Publishers.

Thoumi, F. E. (1999). Las drogas ilegales y relaciones exteriores de Colombia: Una visión desde el exterior [illegal drugs and Colombia's foreign relations: A vision from abroad]. In A. Camacho-Guizado, A. López-Restrepo, & F. E. Thoumi (Eds.), *Las Drogas: una Guerra Fallida* (pp. 93–145). Bogotá: Tercer Mundo Editores.

U.N. Drug Control Programme (UNDCP). (2000). *Mapa localización de cultivos ilícitos en colombia, censo 2000* [map of illicit crop localities, 2000 census]. Bogotá: División Nacional de Estupefacientes, Dirección General Antinarcóticos de la Policía Nacional and UNDCP.

U.N. Office of Drugs and Crime (UN-ODC). (2003). *Global Illicit Drug Trends 2003.* United Nations Publication E 03 XI 5. New York: United Nations.

U.N. Office for Drug Control and Crime Prevention (UN-ODCCP). (1999). *Global Illicit Drug Trends 1999.* United Nations Publication E 99 XI 15. New York: United Nations.

U.N. Office for Drug Control and Crime Prevention (UN-ODCCP). (2000). W*orld Drug Report 2000.* Oxford: Oxford University Press.

U.N. Office for Drug Control and Crime Prevention (UN-ODCCP). (2001). *Global Illicit Drug Trends 2001.* United Nations Publication E 01 X 11. New York: United Nations.

United States Department of State (USDS). (2001). *The Andes Under Siege: Environmental Consequences of the Drug Trade.* http://usinfo.state.gov/products/pubs/andes/.

Vargas, R. (1999). *Fumigación y Conflicto: Políticas Antidroga y Deslegitimación del Estado en Colombia* [Fumigation and conflict: drug war policies and the loss of legitimacy of the state in Colombia]. Bogotá: Tercer Mundo Editores.

Vásquez, R. (1997). La agricultura Colombiana en 1996 [Colombian agriculture in 1996]. *Agronomía Colombiana, 14*(2), 158–181.

Velaidez, R. (2000). Impacto de los cultivos ilícitos y las fumigaciones aéreas con glifosato sobre el medio ambiente [impact of illicit crops and aerial fumigation using glyphosate on the environment]. In *Cultivos Ilícitos en Colombia, Memorias del Foro: "Cultivos Ilícitos en Colombia"* 17 y 18 de Agosto, 2000, (pp. 149–155). Bogotá: Ediciones Uniandes.

Voss, R. S., & Emmons L. H. (1996). Mammalian diversity in neotropical lowland rainforests: A preliminary assessment. *Bulletin of the American Museum of Natural History, 230*, 1–115.

World Bank. (1999). *Violence in Colombia: Building Sustainable Peace and Social Capital.* Report No. 1865-CO. Washington, D.C.: The World Bank.

Young, K. R. (1996). Threats to biological diversity caused by *Coca*/cocaine deforestation. *Environmental Conservation, 23*(1), 7–15.

Zuluaga, J. (1999). Sueños de paz. realidades de guerra [dreams of peace. Realities of war]. In L. A. Restrepo (ed.), *Síntesis '99. Anuario social, político y económico de Colombia* (pp. 45–54). Bogotá: IEPRI, Fundación Social, TM Editores.

CHAPTER 9

DEFOLIATION DURING THE VIETNAM WAR

GORO NAKAMURA

1 INTRODUCTION

The rain forests of Vietnam are deep and dark. The dense foliage overhead blocks out the sun. At the base of gigantic tropical trees are colonies of ferns and vines. There is no breeze and the stagnant air is sweltering. The forests become infested with leeches during the monsoon: leaves harbor insidious green leeches while on the laterite earth are ones of limpid amber. In swamps these bloodsuckers are often black.

For the guerrillas, who fought in the Vietnam War it was the very forest that was their stronghold. In it they could find edible nuts, wild taro (*Colocasia antiquorum*) and other necessities of life. Yet it was a dreadful place for American soldiers who possessed no knowledge of tropical forests. If a pilot landed in the forest after his plane was shot down, he would often find himself in great misfortune. His chances of survival would be low.

The United States forces entered the same forest to sweep out the enemy. But once the enemies hid there, it was easy to lose track of them. Dogged pursuit fell into ambush and heavy firearms had no significant affect against an invisible foe. The proposed solutions was to strip the forest bare in order to dismantle the enemy's sanctuary and by razing cultivated fields and restricting farmers to US-controlled territory, leave the enemy with no food (Nakamura, 1995).

Operation Ranch Hand—the Air Force and Herbicides in Southeast Asia 1961–1971, a document issued by the Office of Air Force History, describes how the United States implemented its strategy in the Boi Loi area, Tay Ninh Province:

> Riot gas was dropped on hamlets in the defoliation target area to add to the occupants' incentive to leave. Some of the bombs dropped by the fighters used time delay fuses, and refugees later reported that they were very effective in inducing fear because they exploded at times when no aircraft were present.
>
> (Buckingham, 1982).

For 10 years from 1961 to 1971, the United States ran a defoliation campaign in Vietnam. Its official name was "Operation Ranch Hand", and it is the main subject of this chapter. The chapter will first summarize previous cases of herbicide use in armed

W. de Jong, et al. (eds.), Extreme Conflict and Tropical Forests, 149–158.

conflicts. The main focus, however, are a detailed description of Operation Ranch Hand, its environmental impact, and the health impact of the toxic dioxins that entered the Vietnam environment because of the use of defoliants.

2 ANTECEDENTS OF DEFOLIATION PROGRAMS

The Vietnam defoliation programs were not the first example of the use of chemical warfare in history. The Italian armed forces sprayed herbicides from the air in Ethiopia in 1936. The British also sprayed herbicides in military action against riots in the forest in Malaya during the early 1950s. In World War II, the United States made plans for spraying missions in Japan, but in the end the mission was canceled. The methods were tested, however, in trials at Bayport and Marathon in Florida in 1944. The planned herbicide spraying in Japan towards the end of World War II, although it did not actually take place, shows the seriousness with which the option was contemplated long before it started in Vietnam.

According to historian Barton J. Bernstein (1987), the plan for the use of herbicides against Japan was conceived in May 1945, when the United States Army envisioned the use of the crop defoliant ammonium thiocyanate to devastate Japan's rice production. The proposed targets were rice paddies in the vicinity of major cities including Tokyo, Yokohama, Osaka, Nagoya, Kyoto, and Kobe. General Arnold, commander of the Air Force maintained that bombing industries and cities would be more effective. The development of bacteriological weapons using anthrax and botulin was also under way (Bernstein, 1987). "Truman insisted that he would be ready for biological and gas warfare if the war was to continue for too long. Had it been so, the consequences would have been no less horrible than the dropping of the atomic bombs," said Bernstein in my interview at his room in Stanford University. "Though I can't conjecture the result of B-29 spraying," he added, "if the war had continued till November that year, the defoliation program would surely have become a reality." Three days before the first atomic bomb was dropped in Hiroshima, the commander's deputy of the United States Armed Forces pushed ahead with the planned herbicide use against Japan. A full-fledged scheme emerged only the day after the dropping of the second atomic bomb on Nagasaki, shortly before Japan's surrender (Bernstein, 1987). The project's abandonment was a close call.

3 OPERATION RANCH HAND

Operation Ranch Hand was initially directed by President Kennedy. Annihilation of the natural environment and food supplies in South Vietnam would, Kennedy thought, reduce Viet Cong mobility. How the government would anticipate and dodge the probable public outrage still remained a problem.

The first defoliation test flight was conducted along a road north of Kontum in South Vietnam on August 10, 1961. The operation used commercial planes and CIA pilots for disguise. The Joint Chiefs of Staff noted (Buckingham, 1982):

. . . In conducting aerial defoliant operations . . . care must be taken to assure that the United States does not become the target for charges of employing chemical or biological warfare. International repercussions against the United States could be most serious. In this connection, it is recommended that the operations be covered concurrently with a publicity campaign as outlined by Task Force Vietnam in Saigon.

Soon after the first test flight, volunteers were being recruited for the special aerial spray flight (SASF). The mission required the peculiar conditions that members all must wear civilian clothes, fly aircraft without USAF markings, and not be acknowledged as member of the United States military if they were captured. Those accepted also had to swear not to confide their mission or whereabouts to their families. Even letters home were not permitted. Secretary of Defense McNamara proposed that the program be conducted under South Vietnamese auspices with their markings on the aircraft and a South Vietnamese officer on the crew as the ostensible aircraft commander. He also favored no publicity given to United States participation in defoliation or crop destruction operations. In like manner, American Ambassador Nolting in Saigon recommended that the chemicals for use in Operation Ranch Hand be manifested as civilian cargo, thus warding off inspection of munitions shipments by the International Control Commission (Buckingham, 1982). The Tan Son Nhut Airbase provided a fenced haven for the spray planes, a space that usually housed President Diem's private aircraft. Ranch Hand crews were billeted near the runaway.

Such camouflage maneuvers suggest the United States Army was well aware that it was embarking upon chemical warfare. The United States at that time had not yet ratified the Geneva Accords of 1927 on chemical weaponry. However, it could not afford to make an ostentatious gesture of defiance against the agreement. The project had to appear as if directed by the South Vietnamese government. A Saigon newspaper reported on January 11, 1962 the government statement:

"The Republic of Vietnam today announced plans to conduct an experiment to rid certain key communications routes of thick, tropical vegetation. United States assistance has been sought to aid Vietnamese personnel in this undertaking. The purpose of this operation is to improve the country's economy by permitting free communications along these routes and by making additional land available for cultivation and other uses. The (Vietnamese) Government emphasized that neither of the two chemicals is toxic, and that neither will harm wildlife, domestic animals, human beings, or the soil"

(Buckingham, 1982; CINCPAC, 1963).

In short, the report promulgated a feigned scenario in which the leading protagonist was the South Vietnam government, with the United States a supporter, and the project's aim was the improvement of the economy.

The South Vietnam government produced a classified document entitled *Khai Quang Program and Provisions, Document 3/5, The Republic of Vietnam Military Chiefs of Staff Section No. 3 Khai Quang* literally means to deforest and to let light in, and accordingly the *Khai Quang* program here signified Operation Ranch Hand. The document, in Vietnamese, was a comprehensive account of the program code-named Operation 202. The details read as if the South Vietnam forces carried out the program, but the document itself was unmistakably a translation from an English original. The objectives of the program were precisely stated:

> "Destroy Viet Cong bases and asylum in the jungle and create a vacuum in which no hideout is possible.... Deprive them of arable land in order to cut off food supplies and to damage their self-sustenance system. The program should encourage among the opposition demoralization as well as distrust of their leaders."

The psychological warfare plan went as follows:

> "Either the distribution of flyers or announcement by loudspeaker is to be used to promote public understanding of the safety of the chemicals. As for mass propaganda, the Psychological Warfare Bureau has provided each strategic post with four million flyers in which are elucidated the advantage and safety of the *Khai Quang* agents and which should be suitable for any region."

As the program proceeded, however, the camouflage did not hold long. Adamant that the defoliants caused no harm to humans or animals, the United States government policy was later to veer toward overt implementation and reinforcement of the program. (RV/MCS, 1969).

The data of Vietnam defoliant and other spraying are recorded in the HERBS file, the United States Air Force computerized record of Ranch Hand missions. The files include detailed information on each flight, such as the date, target, amount of herbicide sprayed, and so on. The HERBS file records in total 6,542 spray missions (Buckingham, 1982). The files, however, recorded only 86% to 88% of the entire program. It is missing spray missions performed by H-34 and UH-1 helicopters, and the spraying around the United States military bases routinely carried out by nonRanch Hand American soldiers. A Veterans Administration report on the HERBS file indicates that 83.6 million liters were sprayed, a rough estimate covering only mission sprayings. Other data show that 1.6 million liters of herbicide were sprayed in Laos from 1965 to 1969 (Buckingham, 1982). Withheld data concerning spray missions in Cambodia would also increase the total (NAS, 1993). An estimated total amount of 91 million liters was probably sprayed during the Vietnam War. Westing (1984) estimates the total amount of sprayed chemicals at 102.5 thousand tonnes, including 90.5 thousand tonnes of herbicides, nine thousand tons of the antipersonnel Agent CS, and 3,000 tonnes of the insecticide malathion.

Twenty-four C-123 provider planes of the 12th Air Commando Squadron flew the spray mission code-named "Trail Dust." Each converted cargo plane was equipped with a one thousand gallon tank and an A/A45Y1 automatic spray system which pumped the defoliants at high pressure from 36 nozzles on both wings and tail. The planes cruised at 40 m height or less with a speed of 240 km/h, and took no more than four minutes to empty a tank of defoliant. Sorties were usually made at dawn while the ground was still cool and there were no ascending thermals to disperse the dumped chemicals before they settled on the ground. One C-123 was capable of reducing to utter devastation a strip of land 80 m wide by 16 km long in one single flight. Winds often served to spread the chemicals even further. When missions were exposed to gunfire from the ground, they just dumped the whole load at once and took off. A regular sortie consisted of three to six aircraft in formation, and during 1967 to 1969 when the activity reached its peak, the spray unit flew 18 to 27 sorties a day. Forests that were first sprayed with chemicals and then bombed with napalm or oil fires became wastelands. Starting with the trial spraying on August 10, 1961, the last C-123 sortie of Operation Ranch Hand flew over cropland in Ninh Thuan on January 7, 1971, and ended with a helicopter mission on October 31.

The impact on Vietnam's forest has been devastating. Ten years of spraying left the Vietnamese rainforests in tatters. Large areas of mangrove and rainforest were eradicated and cropland devastated. The summary tally is that 104,909 ha of mangrove forests and about 3,000,000 ha of inland forests were eradicated, and 202,000 ha of croplands devastated. The total area defoliated amounted to 3,104,000 ha. Some 82,830,000 cubic meters of wood were lost. Inland forests were converted into the fields of tall grass whose strong, thick reticulate roots resist the farmer's hoe (Nakamura, 1995).

4 A CASE STUDY: CA MAU PENINSULA

Fifty-five missions were flown over Ca Mau peninsula between 1966 and 1970 and they destroyed most of its mangrove forest. One hectare of thick mangrove forest contains roughly 350 trees, with an average height of 26 m and a timber volume of about 100 m^3. Mangrove forests make up 80% of the 444,918 ha of forests destroyed in Ca Mau (Nakamura, 1983, p. 83). Visiting the sprayed area more than 20 times, I could confirm that the mangrove and brackish water forests were transformed into a sterile desert. The intense climate, including the torrential monsoon rains and the strong winds of the dry season completely denuded the area and eroded the fertile soil, leaving now only flat, empty, barren land. Ecologists estimated that a forest destroyed this way will take at least a century to recover. Now, more than 30 years after the cessation of the defoliation project, Cape Ca Mau shows no signs of recovery. It seems that the forest destroyed, not by a natural phenomenon such as a forest fire or harsh weather, but by intensive chemical spraying, takes longer to regenerate.

The chemical components of herbicides decompose within a few months after spraying. However, the dioxins[1] that are present as impurities in defoliants like Agent Orange stay exceptionally well intact in the soil and are eventually dispersed by natural agents over a broader area. So it is quite possible that these chemicals have also affected subterranean organisms, resulting in some alteration in soil regeneration and the quality of the soil.

The mangrove tree, a member of the Rhizophoraceae family, grows where there is a confluence of fresh water and seawater. It produces abundant seeds that drop into the water and are carried by currents and tides to the area where they germinate.

---

[1] Dioxin refers to a group of compounds characterized by two benzene rings joined by a double oxygen bridge to which chlorine atoms are attached. The toxicity depends on the position of chlorine atoms on the benzene rings.

Rhizophora apiculata, which grows 20 to 30 m high, used to be the dominant mangrove species in the Ca Mau forest, followed by the 10 m tall *Bruguiera parviflora* and *Avicennia alba*. But herbicides have killed all individuals of these species. Now you see only shorter tree species such as *Excoecaria agallocha* and *Ceriops tagal* that measure around 8 m when full grown. Ha Quoc Hung of the Ca Mau forestry office, a young Vietnamese man who had studied forestry at a university explained: "We cannot just rely on nature's ability for recovery; an alternative option is to plant mangrove seeds during the appropriate season." Vietnam's mangrove planting program is expected to rehabilitate over 160 thousand ha of destroyed forest land. The patient rural dwellers of Ca Mau already started this project and plant mangrove trees one seed at a time, repeating the effort the next year if the seed failed to germinate and grow. The Ministry of Agriculture and Rural Development of Vietnam has announced that more than 70% of destroyed saltwater areas had been replanted to mangrove trees as of the year 2000.

The destruction of the forests from defoliants also affected wildlife. Many species of monkeys, birds, alligators, and snakes were destroyed or migrated to other regions. For a while after the defoliation of the Ca Mau peninsula the numbers of fish and prawn in the waterways increased due to the large amount of leaves that dropped in the water, enriching the food sources of these populations. After 2 or 3 years, however, many kinds of fish had disappeared from the affected area.

Among migratory birds that came to Japan every year were egrets from the Ca Mau wildlife sanctuary. Noda in Saitama prefecture, north of Tokyo, had a remarkable colony of such egrets. In the late sixties, however, the number of egrets coming to Noda began to drop sharply. People wondered what had happened. At that time farmers had started to use more chemicals and that had driven the fish out of the rice paddies, the fish that are the main part of the egrets' diet. People concluded that the egrets had left because they lost their main source of food. This guess was only partially right, for egrets had also suffered from herbicides in their homeland, Vietnam. Many egrets and their eggs perished in the Ca Mau sanctuary.

5 MASSIVE ESCALATION

American chemical manufacturers, such as Dow, Monsanto, Hercules and Diamond Shamrock, were mobilized for the production of the herbicides. From the time the United States launched defoliation missions in eastern and southern Laos in December 1965 to reinforce attacks on the Ho Chi Minh Trail, frequent shortages of herbicides occurred. Even the drastic measure of appropriating almost all domestic production of herbicides for Operation Ranch Hand did not ease matters. With the increased sprayings of 1967, demand for herbicides reached a peak. The April 1967 issue of *Business Week* reported that the amount of 2,4,5-T demanded by military authorities was three times domestic production. In October the same year in Japan, Mitsui Toatsu Chemical in Omuta "had begun, on foreign demand, manufacturing 2,4,5-TCP, precursor to the organic chlorine herbicide 2,4,5-T, with an average production of 55 tonnes a month to be exported chiefly to Australia and New Zealand". The manufacturer denied involvement in the Vietnam War, and the question whether their product was eventually passed into the hands of the United States led to polemics in the Japanese Diet in 1969 (minutes of the Committee for Foreign Affairs of the House of Representatives of Japan, July 23, 1969; *Asahi Shinbun*, July 24, 1969; Nakamura, 1995).

6 DIOXIN CONTAMINATION

Agent Orange made up 67% of the chemicals sprayed in the war. The name Agent Orange came from the color code painted on the drum, not the color of the fluid itself. Agent Orange is an even mixture of two organic chlorine herbicides: 50% 2,4,5-T (2,4,5-trichlorophenoxyacetic acid) and 50% 2,4-D (2,4-dichlorophenoxyacetic acid). Other 2,4,5-T based herbicides of different composition included Dinoxol and Trinoxol. Agent White and Picloram are two other solutions of 2,4-D. Arsenic laced Agent Blue, a lethal herbicide of cacodylic acid was used mainly against preharvest crops.

Agent Orange was used to defoliate the forest of Vietnam. Its use was particularly controversial because one of its constituents, 2,4,5-T, contains the dioxin TCDD (2,3,7,8-tetrachlorodibenzo-para-dioxin), a substance highly toxic for humans. The United States military did not intentionally concoct this mixture. The dioxin is an inevitable impurity produced in the manufacture of 2,4,5-T.

Unlike the herbicide components, 2,4,5-T and 2,4-D which disintegrate in a few months, dioxin cannot be reduced through natural processes. Although ultra-violet light degrades dioxin in lab experiments, dioxin already deposited in the soil or absorbed by living creatures is not affected by this radiation. Dioxin hardly dissolves in water but clings to particles of earth and dust. Once settled on the surface of the ground, it mingles with mud and is washed into rivers and the ocean by rain. Plankton take it up and pass it on to small fish as they eat the plankton, and so forth. The concentration of dioxin in living organisms increases through the food chain. Through food and water, the fat soluble dioxin infiltrates into the human body, and accumulates in fatty tissues where it will linger for years.

TCDD surpasses other substances in both acute and chronic toxicity. Very small amounts, in the order of one part per trillion (ppt), or even one part per quadrillion (ppq), can be a threat to humans. Agent Orange and other herbicides used in the Vietnam War contained some 1 to 50 ppm of TCDD, implying that the whole of the South Vietnam became poisoned with TCDD, according to United States Air Force estimate.

7 THE IMPACT ON HUMAN HEALTH

TCDD is carcinogenic and an exposed mother may pass the toxicity on to her fetus. The capacity of dioxin to damage the human reproductive system can lead to congenital abnormalities in children born of an exposed father. A secret Dow Chemical conference in 1965 acquainted the United States government with the toxic effects of TCDD, yet herbicide spraying drastically increased thereafter. The Joint Chiefs of Staff simply filed away the 1966 Rand Corporation's field study, which warned that the herbicides were highly hazardous to Vietnamese peasants and Operation Ranch Hand servicemen.

The United States pushed ahead with the chemical warfare in the name of "protection for American soldiers."

In the chapter on the use of antiguerrilla weaponry, the training textbook TC3-16, issued by the Department of the Army in April 1969, stipulates the safety of Agent Orange for humans and animals, explaining that there had been no reports of adverse effects among those exposed to herbicides in Operation Ranch Hand. It also contained passages, however, that implied the toxicity of the chemicals; that is, although no precautions are necessary for servicemen who are exposed to the agents through handling them, they should shower and change clothes afterwards, and soiled spray planes must be washed with soap till the chemicals are completely removed (JAASC, 1969).

In June 1969, the Saigon newspaper, *Tin Sang*, began a report series on the disproportionate increase in abnormal births in South Vietnam that seemed associated with the defoliation program. The series was quickly censored after it started, but already in the United States momentum against Operation Ranch Hand was developing. K. Diane Courtney submitted to the National Institutes of Health a study confirming that 2,4,5-T caused birth defects and stillbirths among mice. A year after the *Tin Sang* report was smothered Harvard scientist Matthew S. Meselson and his research team went to Vietnam and warned strongly against Operation Ranch Hand. Another follow-up report informed that the concentration of dioxin in Agent Orange was many times the permissible level for domestic herbicides in the United States. Pushed by growing public concern for the safety of Ranch Hand servicemen, the Nixon administration at last began edging toward the diminution of the program. Still, complete cessation of the program only came late in 1971.

Although memories of Operation Ranch Hand have faded into the past, what it sowed has never faded or failed. Soldiers have returned to their country only to resume a long, tough battle—this time against a new invisible enemy called toxic dioxin lurking inside their bodies. In Vietnam, the Vietnamese people have had to endure the destruction and pollution of the 10-year-long spraying and its aftermath.

The United States made several miscalculations regarding the use of chemical weapons in the Vietnam War. The contamination of herbicides among their own servicemen exceeded what had been expected. As long as the US military sprayed from the air, American soldiers could not have been exposed to dioxin, or so they thought. However, the reality in the battlefields was different.

"We entered the jungles and kept up sweeping operations after the herbicides were sprayed, because we didn't know the Air Force sprayed the chemical," said Daniel J. Loney in 1982 at his home in Philadelphia. He had been sent to Vietnam at the end of 1966 as a soldier of the 173 Airborne Brigade and joined the battle of Boi Loi Woods, in the Parrot Beak area of Tay Ninh, in the Iron Triangle of the Central Highlands and Laos in 1967 to 1968, and now lives in Philadelphia. He recalled, "We wore the same clothes soaked with the herbicides for weeks. We fought in Vietnam drinking the contaminated river water." In 1968, he had an attack of high fever and was sent back from the battlefront to a field hospital in Cam Ranh. From there he was discharged from military service. After he returned home, he felt numbness in his limbs and a rash developed on his skin. In 1980, his daughter was born with congenital deformation on her right arm (Nakamura, 1983).

The level of impact was exacerbated because no one was afraid of the exposure, a result of the widely disseminated propaganda that the chemicals were harmless. Because the United States lost the war they could not establish an "Agent Orange Casualty Commission" in Vietnam. Unlike the atomic bombing in Japan where the United Sates was able to examine survivors in Hiroshima, it had to examine the effects of dioxins by studying its own veterans instead of the Vietnamese people. A study in *the Journal of the American Medical Association* (Anonymous, 1988) showed that the amount of dioxin in fat tissue in Vietnam veterans reached 41.7 ppt in average, and in blood it was 46.3 ppt. This was 10 times higher than among the general public of the United States with an average of 4.3 ppt.

In July 1993, the National Academy of Science (NAS) submitted a report on the health risk of dioxins. The Committee to Review the Health Effects in Vietnam Veterans of Exposure to Herbicides at the Institute of Medicine, NAS reviewed 6,420 scientific and medical articles regarding the effects of dioxin exposure on the human body, and thoroughly examined more than 230 significant studies. The committee confirmed the association between herbicide exposure and five diseases, including three types of cancers: soft tissue sarcoma, non-Hodgkin's lymphoma, and Hodgkin's disease, and two types of skin disease: chloracne and porphyria cutanea tarda. In addition, three additional diseases: lung cancer, prostate cancer, and multiple myeloma were undeniably associated with herbicide exposure. The Federal Government and Department of Veterans Affairs soon decided to add these to the list of diseases subject to relief measures for Agent Orange victims, and to pay compensation for nine diseases including chloracne. In 1996, they added spina bifida, as limited evidence suggested an association between herbicides and this congenital deformity of veteran's children. Acute myelogenous leukemia was also added to the list in 2000 (IM/DHPDP, 2000).

The number of veterans who have developed or will develop symptoms is estimated to be 10% of the approximately 2,590,000 men and women who served in the Vietnam War. As of 1995, 99,226 veterans had become aware of their symptoms, and they have been receiving medical treatment.

8 CONCLUSION

Operation Ranch Hand in Vietnam ended in 1971, 10 years after it began. The Vietnam War lasted 15 years and ended with the fall of Saigon in 1975. In postwar Vietnam, ruined land was left as if it was in exchange for peace. Regardless whether former ally or enemy, every Vietnamese has carried the burden of the war. There is no safe place to escape from dioxins for those who live in South Vietnam.

More than a quarter century has passed since the War ended. The dead trees caused by Operation Ranch Hand are gone, but the problem remains how to revive ecosystems and woodlands. Since then the Convention on the Prohibition of the Development, Production, Stockpiling and Use of Chemical Weapons and on their Destruction has been adopted by the United Nations General Assembly in 1992. The Convention includes a ban on the use of herbicides as a weapon in war, and it has been ratified by 143 nations now.

REFERENCES

Anonymous. (1988). Serum 2,3,7,8-tetrachlorodibenzo-p-dioxin levels in US army Vietnam-Era veterans. *Journal of the American Medical Association,* 260 (Sept 2).

Bernstein, B. J. (1987). The birth of the United States biological-warfare program. *Scientific American,* (June), 120–121.

Buckingham, W. A. (1982). *Operation Ranch Hand—The Air Force and Herbicides in South East Asia 1961–71.* Washington, DC.: Office of Air Force History, United States Air Force.

CINCPAC. (1963). *The Psychological Aspects of the Use of Defoliants in the Republic of Vietnam. Report to the Joint Chiefs of Staff (JCS),* Mar 22, p. A-1.

Institute of Medicine, Division of Health Promotion and Disease Prevention (IM/DHPDP). (2000). *Veterans and Agent Orange: Update 2000.* Washington D.C.: Committee to Review the Health Effects in Vietnam Veterans of Exposure to Herbicides National Academy of Sciences.

Japan A-A Solidarity Committee (JAASC). (1969). *Sensou Hanzai [War Criminal]: Training Circular No. 3–16, Employment of Riot Control Agents, Flame, Smoke, Anti-plant Agents and personnel Detectors in Counter-guerrilla Operations.* Washington D.C.: Department of the Army, April 9, 1969, Chapter 5-1-(4), 62. 193. Tokyo: Rodo Jumpousha.

Nakamura, G. (1983). *Haha wa Karehazai wo Abita* [Mothers Were Exposed to Agent Orange]. Tokyo: Shincho sha, 129.

Nakamura, G. (1995). *Senjo no Karehazai* [Agent Orange in the Vietnam War—Vietnam, America and Korea]. Tokyo: Iwanami Shoten.

National Academy of Sciences (NAS). (1993). Veterans and Agent Orange. Washington, D.C.

RV/MCS (Republic of Vietnam, Military Chiefs of Staff, Sec.3). (1969). *Khai Quang Program and Provisions, Document 3/5.* Hanoi.

RV/MCS (Republic of Vietnam, Military Chiefs of Staff, Sec.3). (1984). *Herbicides in War–The Long-term Ecological and Human Consequences.* Stockholm: Stockholm International Peace Research Institute (SIPRI).

CHAPTER 10

ADDRESSING EXTREME CONFLICTS THROUGH PEACE PARKS

JEFFREY A. MCNEELY

1 INTRODUCTION

The end the Cold War has not brought violent conflict to an end. The war in Afghanistan, the war in Iraq, as well as recent events in New York, London, Madrid and Bali suggest that the danger of potential destructive wars remains high. Civil wars or other smaller scale conflicts continue in many African countries, killing and displacing vast amounts of people from the Sudan to the Congo. Drug cartels continue to wreak havoc in countries in South America. Similar examples can be given from Cambodia, Myanmar, Philippines, Indonesia, Nepal and Sri Lanka. These are examples of conflict that affect societies at large in many tropical forest countries. Because of modern means of communication, increasing demand on resources driven by higher consumption levels and larger number of people, and the spread of modern instruments of war, future wars are likely to be extremely destructive for both people and nature.

The convention on biological diversity (CBD), finalized in 1992 and now ratified by 189 countries, commits signatories to conserve biodiversity within the boundaries of their territories (Article 1). They also pledge that activities permitted within their boundaries or control will not damage the environment (Article 3). However, many states support or do not deal adequately with these conflicts and, in fact, could be held responsible for the negative impacts on biodiversity under the CBD. But before states can be called upon to address problems of war or other types of violent conflict and their impacts on biodiversity, it is necessary to understand better what are the impacts of war, preparations for war, and managing the aftermath of war on biodiversity. This chapter proposes some key issues that need to be considered when trying to understand these impacts.

Before going into the discussion on what impacts modern wars have on biodiversity, the chapter will dwell briefly on how avoidable wars can be, and therefore, how avoidable negative impacts on biodiversity can be. The International Commission on Peace and Food (1994) suggests that armed conflicts or revolutions have been a key factor in many international political and security changes. Not only has some war defined most modern societies, the possibility of war has been a principal organizing

W. de Jong, et al. (eds.), Extreme Conflict and Tropical Forests, 159–172.

factor and political stabilizer. This suggests that a lack of ability to adapt to change eventually leads to maladaptation. This maladaptation is eventually being corrected through war (Edgerton, 1992).

One intermediate factor that links a state's maladaptation to war is resource scarcity. Homer-Dixon (1994) argues that resource scarcity is one of the primary factors that has explained many previous wars and is likely to lead to new wars in the future. Scarcity conflicts over land, water, forest, oil and other essential natural resources are likely to increase. New wars, or other kinds of violent conflicts, will occur unless major efforts are taken to prevent them. While many would probably agree that adaptation or adequate prevention is possible through the application of knowledge and wisdom, history does not support such optimistic views. War has been a common phenomenon in human societies as a means of resolving conflicts arising from various sources of resource scarcity (Keeley, 1996). Biodiversity-related problems like desertification, soil erosion, deforestation, and water scarcity reduce food-growing potential, worsen health effects, and diminish life-support capacity. This can lead to civil conflict and increase the likelihood of war. Nietschmann (1990) concludes on the basis of experience from Nicaragua, that degraded land and resources are reasons for taking up arms as much as are repression, invasion, and ideology.

The objectives of the Convention on Biological Diversity advocate for the sustainable use of biological resources, and sharing the benefits of such use in a fair and equitable manner. These objectives link the convention squarely with the prevention of future wars. These objectives, therefore, can become critical elements in discussions of national security in tropical forest countries. Investments in sustainable forestry, conservation of water resources, land reform, and the conservation of protected areas can become significant contributions to peace.

This chapter begins by assessing some of the negative impacts of war and postwar times on tropical forest biodiversity, but it will also suggest some of the positive impacts. The chapter will then briefly discuss how war functioned as a traditional social means that human societies used to adapt to changing environmental conditions, and how modern emerging scarcities may fit in this general pattern. It will conclude by showing how conserving biodiversity can contribute to peace, building on the preamble to the Convention on Biological Diversity, which states that: "Ultimately, the conservation and sustainable use of biological diversity will strengthen friendly relations among states and contribute to peace for humankind."

2 THE IMPACTS OF WAR ON BIODIVERSITY IN TROPICAL FORESTS

2.1 Negative Impacts of War on Biodiversity

Where war occurs, the collective actions of large numbers of mostly male fighters cause negative impacts as conflicts become an excuse to ignore restraints of practices that cause environmental damage on biodiversity in tropical forests. War, and preparations for it, may have negative impacts on different levels of biodiversity. The impacts can be direct through hunting and habitat destruction by armies. It can also be indirect, for example through the activities of refugees who flee the battle areas, concentrate often in forest-rich areas and harvest resources from their new environment (often to the consternation of resident communities).

The impacts on biodiversity from war are often deliberate. It has even led to a new term in the military vocabulary, "ecocide," a term that means the destruction of the environment for military purposes. Although the word is new, the practice is ancient, and sometimes called the "scorched earth" strategy. Westing (1976) makes a distinction between two types of ecocide. One is meant to deny to the enemy any habitats that produce food, refuge, cover, training grounds, and staging areas for attacks and involves massive and extended disruption of elements in the human and natural landscape. The second one is aimed at releasing dangerous forces through planting landmines or releasing microorganisms that cause disease (Strada, 1996).

Mentioning only a few of the many examples that could be given here suffice to illustrate the point. In Vietnam, US forces cleared millions of hectares of forest by spraying herbicides to reduce ambushes (Westing, 1982; see also Chapter 9). The impact of this modern ecocide was severe. Ten percent of the country was spread with herbicides, affecting 50% of the mangroves in Vietnam. It replaced high-diversity forests with grasslands, highly productive mangroves with mudflats, and decreased both freshwater and coastal fisheries (Nietschmann, 1990). A more modern equivalent is the intensive logging to fund the war that the Karen and other tribal groups are fighting with the military government of Myanmar. These tribal groups inhabit densely forested mountain regions along the country's borders with Bangladesh, India, China, Laos, and Thailand. While it benefits their cause, this logging at the same time destroys forest cover and makes the guerrilla groups more vulnerable to attacks (Harbinson, 1992). The general lawlessness along the border with Thailand has greatly increased the flow of logs, both with and without government permission, leading to the virtual clear felling of many of the country's most productive forests.

Recent violent events in Africa have caused true disasters for biodiversity in tropical forests. The rare and endangered mountain gorilla (*Gorilla gorilla*) inhabits the Virunga Volcanoes region, which forms part of the Central African countries of Rwanda, Democratic Republic of Congo, and Uganda. Only some 600 individuals still remain. From this region Tutsi rebel groups launched their civil war against the government of Rwanda. While the war spread deeper into Rwanda until 1994, it sent large numbers of refugees fleeing to North Kivu District in what is now DRC. Military groups took over headquarters of several tropical forest World Heritage site, including Virunga National Park, Kahuzi-Biega National Park and the Okapi Wildlife Reserve. Some 850,000 refugees lived around Virunga National Park in 1994. Looking for food and firewood, they partly or completely deforested some 300 km^2. Up to 40,000 people were estimated to enter the park every day, harvesting between 410 and 770 tons of forest products. This seriously damaged the bamboo forests, and reduced the populations of elephants, buffalo, and hippos. Organizations such as the Red Cross, Médecins Sans Frontière, and CARE have supported well-meaning relief operations on the park boundaries, but when doing so established dumps for medical wastes inside the park. This caused a serious risk of disease transmission (Pearce, 1994). As a result of the

Table 10.1. *Examples of negative impacts of war on biodiversity or conservation efforts.*

Year	*Event*
	In the Democratic Republic of Congo, civil war stopped efforts to protect the last habitat of the pygmy chimpanzee, a species endemic to that country. Local people hunt this chimpanzee because they increasingly depend on the forest for survival.
1996	In Burundi guerrillas fighting the government used the Kibira and Ruvubu national park as sanctuaries. Government troops started operations in the area and both sides engaged in heavy poaching.
	In India guerrillas from the Bodo tribe took over Manas Wildlife Sanctuary, a World Heritage site, burned down park buildings, looted facilities, killed guards, destroyed bridges, poached rhinos, elephants, tigers, and other wildlife, cleared forests, and depleted fish stocks in the Manas river.
1989	In Sri Lanka, the Wilpattu National Park was attacked by Tamil rebels, killing over a dozen guards and destroying facilities. This caused a withdrawal of conservation staff and an increase in military activity.
	Liberia's rural people, forced by the civil war, hunt duikers, pygmy hippos, forest elephants, and chimpanzees for food.
1983	In Nicaragua the Contras kidnapped the administrator and two rangers of Saslaya National Park (15,000 ha). The National Environment Agency abandoned its management (Thorsell, 1990).
1955–1972	In Sudan the 17 years civil war exterminate the white rhino (Abdullah, 1997).
	In Vietnam during the American war helicopter gunships target elephants because they might be used as pack animals by the Viet Cong.

battle with insurgents, 80 of Virunga's park staff have been killed since 1996. The examples described here, and provided in Table 10.1 lead to a conclusion that is not surprising: War is bad for biodiversity.

2.2 Postconflict Biodiversity Loss

Where war had negative impacts on biodiversity, this situation is not easily reversed when peace returns. For instance, in the 1960s, when Indonesia and Malaysia were fighting over border claims on the island of Borneo, they did relatively little damage to its vast wilderness. Since then, and after fighting had stopped, both countries peacefully competed to cut down and sell their forests. This has not only led to high impact logging. Forest fires that caused damage estimated at USD 4.4 billion have occurred in these logged forests. This has been followed by clearing for industrial timber and oil palm plantations. The prices of these commodities have fallen considerably in recent years, making the damage done even more a waste of both investments and biological resources.

Similar postconflict damage resulted from the Indochina war in Cambodia. The fighting left the country without effective government or rule of law. As a result the country's forest and fish resources turned into a virtually unprotected treasure that was hijacked by those who controlled guns. Both were depleted according to World Bank and Asian Development Bank studies. Uncontrolled logging, much of it illegal, could clear virtually all economically productive forests in the country within 5 years. Export oriented fishing from Cambodia's Tonle Sap (Great Lake) is leading to over-harvesting. Even worse, the ecological productivity of the lake was based largely on the 10,000 km^2 of flooded forest that ensured a healthy flow of nutrients into the lake. Military commanders have come to regard the forest resources as a supplemental source of finance since 1993. Now, less than 40% of the flooded forest remains under natural vegetation. This affects the climate, causes erosion that fills irrigation channels and fishing grounds with silt, and leaves Cambodian farmland more vulnerable to both drought and flooding. Ironically, this complex of problems is very similar to that which Cambodia faced some 400 years ago, when the great civilization centered on Ankor Wat collapsed under environmental pressure (McNeely & Wachtel, 1988).

Vietnam's forests are under greater pressure now that peace has arrived than they ever were during the country's wars; Nicaragua's forests are now under renewed development pressures; and Laos is paying at least part of its war debts to China and Vietnam with timber concessions. I was told in Laos that the Chinese and Vietnamese timber merchants and logging companies are able to operate with impunity in Laos, irrespective of logging regulations, protected area boundaries, or any other considerations. This is perhaps not surprising given the dependence of the Pathet Lao on the support of Vietnam and China during the Indo-China wars (see Chapter 5 by Fujita *et al.*). The motivations may be nobler in times of peace, but the impacts of inappropriate development on biodiversity often are even worse than the impacts of war. Market forces may be more destructive than military forces, while the latter may moderate the former.

2.3 *War Protecting Biodiversity*

But war, or the threat of war, can also have a completely opposite impact on biodiversity, at least in some places and in some conditions. Where combatant armies effectively designate war zones as "off limits" to casual wanderers, wildlife can benefit. War has quarantined large areas of Africa from hunters and poachers (Myers, 1979). The benefits that war may provide to biodiversity are not planned side-effects, but incidental, inadvertent, and accidental. But perhaps the positive impact of war on biodiversity provides some support for the view that war helps societies adapt to their dynamic environmental conditions. This justifies reviewing some of the cases where war, or preparations for war, has benefited biodiversity.

To start again with the Vietnam War, while most of it was a biological disaster, as explained before, it did provide an important contribution to conservation of species. During the war the Migratory Animals Pathological Survey did extensive, long-term research on migratory birds in eastern Asia (McClure, 1974). The research was justified by its relevance to the war effort but the data are now extremely useful for numerous civilian conservation applications. The Annamite Mountains, down which the Ho Chi Minh trail ran, was one of the most heavily-bombed parts of the Vietnam War. In the mountains, new discoveries of previously unknown species have recently been made. Two species of barking deer (*Megamuntiacus vuquangensis* and

Muntiacus truongsonensis), a unique variety of forest antelope (*Pseudoryx nghetinhensis)*, and a bovid ultimately related to wild cattle (*Pseudonovibos spiralis*) have been found here (Dillon & Wikramanyake, 1997). In addition a species of pig that formerly was known only by a few fragmentary specimens was rediscovered. The survival of such species in such a heavily-bombed area is testimony of the recuperative power of nature. It also shows that wildlife may be able to withstand even the most extreme kinds of human pressure during warfare. Now, more than three decades after the war ended, these species are more threatened by development activities than they were by the Indochina wars.

Examples can be found in other countries of Asia. During the mid-1960s to mid-1970s insurgents operated along the border between Thailand and Peninsular Malaysia. The Belom Forest Reserve on the Malaysian side of the border was closed off by the military, and some 160,000 ha remained untouched by modern logging pressures. Now this area remains rich in wildlife, because of which Malaysia is now converting it into a national park. Where once insurgents operated, now a transboundary protected area exists with matching protected areas in southern Thailand.

Similar examples can be quoted for Africa. For example, Fairhead and Leach (1995) report that parts of the Ziama region of Guinea, which includes an extensive biosphere reserve, became forested following a series of wars that affected the area from 1870 to 1910. The resident Toma people first fought with Mandinka groups from the north and subsequently with the French colonial armies, causing major depopulation and economic devastation that in turn allowed the forest to reclaim agricultural land. The human disaster of war enabled nature to recover.

In Nicaragua the 20 years of fighting allowed biodiversity to recover from a long history of exploitation. Trade in timber, fish, minerals, and wildlife that thrived before the conflict, was sharply reduced. The number of cattle in the country, roughly the same as human population when the war started, declined to two-thirds. Forest could recover on pasture, which led to the recovery of animal populations such as white-tailed deer (*Odocoileus virginianus)*, collared peccaries (*Tayassu angulatus*), mantled howler monkey (*Alouatta villosa*), white throated capuchin (*Cebus capucinus*), night monkey (*Aotus paniscus*), red backed squirrel monkey (*Saimiri oerstedii*), crocodiles (*Caiman crocodilus)*, iguanas (*Iguana iguana)*, large birds, and various mammalian predators. The war destroyed fishing boats, leading to significant declines in shrimp, lobster and other kinds of fishing. This restored various marine populations.

Whether the final balance of the impact of war on biodiversity could be positive in any example may be questionable, however (see Table 10.2). In Nicaragua, hunting by soldiers had at least local negative impacts on wildlife. New military bases and roads were established in formerly remote areas, opening them up to exploitation. The country's once outstanding system of protected areas fell into neglect, and new areas planned were not established. The collapsing economy forced villagers into environmentally destructive activities, including clearing forest for firewood and harvesting wildlife for food. Biodiversity rejuvenated by the war came under renewed threat by people impoverished by the war. The postwar period saw a great acceleration of such impacts and now that peace has broken out, biodiversity is under renewed pressure.

Table 10.2. *Negative and positive impacts of war on biodiversity.*

Negative impacts	*Positive impacts*
• Deforestation	• Creates "no-go" zones
• Erosion	• Slows or stops developments leading to biodiversity loss
• Wildlife poaching	• Focuses state resolve
• Habitat destruction	• Reduces pressure on some habitats
• Pollution of land and water	• Allows vegetation to recover in some areas
• Reduces funds for conservation	• Disarms rural populations, thereby reducing hunting
• Stops conservation projects	• Can increase biodiversity-related research
• Forces people on to marginal lands	
• Creates refugees who destroy biodiversity	

3 BIODIVERSITY LOSS AS A CONTRIBUTOR TO CONFLICT IN TROPICAL FORESTS

3.1 The History of War and Biodiversity

Long-term interactions between people and their environments reaching back at least as far as the origins of fire have considerably influenced today's biodiversity (see, for example, Flannery, 1994; Martin & Klein, 1984; McNeely, 1994; Ponting, 1992). The greatest diversity of terrestrial species today is found in forested areas inhabited by tribal and other indigenous peoples. In these locations relatively large areas of forest serve as buffer zones between communities that historically may have been embroiled in constant warfare, including sneak attacks, revenge killings, kidnappings, and raids on livestock (Keeley, 1996). Various forms of war have been part of the way traditional societies adapted to changing conditions, and helped contribute to the rich biodiversity found today in many tropical forest areas occupied by traditional and indigenous peoples. One could argue then that bringing peace to these regions will remove this means of adaptation, requiring other ways to conserve biodiversity and maintain the capacity to adapt to changing conditions. Ember and Ember (1992) concluded that the number of wars between traditional societies could be forecast by looking at the amount of unpredictable natural disasters and severe food shortages. Under uncertain conditions, people tried to protect themselves by going to war to take resources from enemies. They plundered food stores and gardens of neighboring groups, leaving an enemy facing starvation and rendering large areas of territory at least temporarily uninhabited. Examples of this can be found in the Americas, Polynesia, New Guinea, and Africa.

While wars could serve to provide larger areas of habitat to various species of wildlife, they could also lead to significant increases in the pressure of human population on the remaining wildlife populations. Losses and gains of territory were a very frequent result of warfare among preindustrial societies, leading to dynamic tribal boundaries. These frontiers often were places supporting great diversity of species. Keeley (1996) concludes that: "Even in situations where no territory exchanges hands, active hostilities along a border can lead to development of a no-man's-land, as settlements nearest an enemy move or disperse to escape the effects of persistent raiding. Although such buffer zones could function ecologically as game and timber preserves, they were risky to use even for hunting and wood cutting because small isolated parties or individuals could easily be ambushed in them."

Examples from South America demonstrate the richness in terms of biodiversity, especially in terms of large mammals, of these no-man's-lands. At the time of the first contact with Europeans, large settled villages occupied the land along the major rivers in various parts of the Amazon. These societies often gathered hundreds of men drawn from multiple confederated villages who traveled by canoes to attack their enemies. Different societies ended up being located at significant distances from each other, with large buffer zones separating them. These buffer zones became the refugia for wild game (Ferguson, 1989). As a result, in his first voyage up the Amazon's Ucayali River in 1577, Juan Salenas Deloyola contacted three different language groups that were separated from the next by a distance of about 250 km. Myers (1979) considers this to be an example of a no-man's-land, located between the defended territories of adjacent human groups. While the evidence available at present does not support any particular conclusions about the relationship between ecosystem productivity and war, competition for environmental resources very frequently has been a factor in war between different communities in Amazonia (Ferguson, 1989). Vulnerability to attack may set a threshold on settlement size, or the threat of raids may encourage people to live together to maintain an adequate defensive force.

Warfare between modern and traditional societies has often involved what might be termed "ecological attacks." As one of the most obvious examples of this, the final destruction of the great herds of American bison (*Bison bison),* the foundation of the Plain's Indian life in the USA, closely coincided with the defeat of the Sioux and Cheyenne in the 1870s. Biological warfare was also used, either accidentally or intentionally. Diseases such as smallpox, measles, and influenza had a major impact on the native populations of the Americas, Australia, and the Pacific Islands because they lacked immunity to the "new" diseases. Perhaps more important, the Europeans also brought ecological transformations that disrupted traditional economies and replaced native ecosystems with new agricultural systems that produced more of the goods required by colonists (Crosby, 1986), leading to fundamental—and perhaps permanent—changes in biodiversity.

In the New Guinea Highlands, the Maring people adjusted relationships between population and land, and the alternated build up of pig herds with slaughter for pig feasts were both linked to warfare. Rappaport (1984) suggested that warfare in this part of the world had a self-regulating function of ecological systems including human and pig populations. Warfare maintained both people and pigs below the carrying capacity of the land, given the prevalent technology available. Some of the New Guinea highland cultures have particularly bloody histories. For example, the Mae Enga fought 41 wars for land between 1900 and 1950, of which six resulted in complete routs of the enemy that led to acquisition of new territory from the defeated clan (Meggitt, 1977). Among the Dani people of the New Guinea Highlands, warfare is responsible for almost

30% of mortality (Heider, 1970). Warfare in association with hunting has been well documented among a number of other New Guinea groups, including the Purari, the Kiwai, the Trans-Fly peoples, the Marind-Anim, the Kolopom, the Jacquia, and the Asmat.

While the existence or intensity of warfare in prestate societies is not a simple linear function of population density, population pressure on the land, or protein scarcity, all of these factors are important contributors. It seems reasonable to conclude that ecological pressure works together with cultural and political dispositions toward warfare. The perception of individual or group land scarcity is a function of socio-cultural as well as ecological organization; perceptions of scarcity are often as important as the pattern of rainfall, the numbers of pigs, or the game animals in the forest (Knauft, 1990). Thus the actual warfare carried out by indigenous peoples of the tropical forests have involved numerous factors reinforcing each other, including increasing human population density, related clearance of forests to increase domestic food production, and declining wild food resources. The subsequent population redistribution certainly had profound implications for biodiversity.

These ancient mechanisms to create and maintain space between groups, increasing resource availability and preserving biodiversity, are often not favored by governments. In both Brazil and Venezuela, governments and the military oppose indigenous reserves close to the border. Geopolitical reasons dominate the Brazilian government's and military's concern. The Venezuelan government is concerned that national and international public opinion may be mobilized to advance the human rights of the indigenous groups and to promote eventual self-development and self-determination. Often both countries identify ecologists and anthropologists as part of an international conspiracy to undermine the ability of the government to control the Amazon territory and its native inhabitants (Hill, 1994). The high-level Congress of the Armies of the Americas (CAA) reduces complex social problems into a black and white opposition between "national security" and "terrorist subversion." Those who advocate Indian rights are often linked to subversive organizations (a group that also included feminists and environmentalists). In essence, the CAA created a mythological history of the relationships between indigenous peoples and their land, defining the problems in terms that required military solutions (Hill, 1994) and ignoring the role of indigenous ways of life in maintaining the rich biodiversity of the upper Amazon, and the dependence of the forest-dwelling people on the biological resources of the forest. Such mythologies are not confined to the Amazon only.

3.2 Scarcity Conflicts in Tropical Forests

In modern times resource degradation, including loss of biodiversity, can create scarcities that push people out of the regions where they live. Insufficient supplies of firewood and timber, depleted aquifers, and soil erosion can form a feedback loop of poverty, insecurity, and environmental degradation. As Kane (1995) points out, "Felled trees, for example, no longer anchor soil, which washes away and clogs rivers, and the disrupted flows of water cause further soil erosion and disrupt harvests of fish. In rural areas where people directly depend on the soil and water and forests for sustenance, poverty is essentially an environmental trend. These people are usually cash poor, yet so long as they are natural-resource rich, they can remain at home and prosper. But when people flee poverty they are often fleeing environmental impoverishment—after the top soil blew away or the well ran dry—in places without a rural economy that offers them alternative sources of livelihood."

Resource scarcity can arise from three sources: degradation or depletion of a resource; increasing consumption of the resource (for example, due to population growth or rising per capita resource consumption); and uneven distribution that gives relatively few people disproportionate access to the resource and subjects the rest to scarcity. Resource scarcity can lead to declining agricultural production, economic hardship, migrations of people from areas of environmental stress, and tensions within and among groups. This mélange of factors can contribute to violent conflict (Homer-Dixon, 1994). When resource scarcity reduces the ability of states to meet the needs of their population, dissatisfaction can lead to declining state authority. Sooner or later such discontent nurtures violent collective action.

Homer-Dixon (1994) concludes that: "Within the next 50 years, the planet's human population will probably pass nine billion, and global economic output may quintuple. Largely as a result, scarcities of renewable resources will increase sharply. The total area of high-quality agricultural land will drop, as will the extent of forests and the number of species they contain. Coming generations will also see the widespread depletion and degradation of aquifers, rivers, and other water resources; the decline of many fisheries; and perhaps significant climate change." Resource scarcities in many parts of the developing world are already contributing to violent conflicts that are probably early signs of an upsurge of violence in the coming decades that will be induced or aggravated by scarcity. Poor people in tropical forest countries will be particularly affected because they are less able to buffer themselves from resource scarcities and resulting social crises. These people typically already are suffering acute hardship from shortages of water, forests, and fertile land. A major dilemma is that fast-moving, unpredictable, and complex environmental problems can overwhelm efforts at constructive social reform. Moreover, scarcity can sharply increase demands on key institutions, such as the State, while it simultaneously reduces their capacity to meet those demands. These pressures increase the chance that the State will either disintegrate or become more authoritarian, both of which enhance the likelihood for conflict, even war.

4 CONCLUSIONS AND POSSIBLE SOLUTIONS

National and international security can no longer be conceived in narrow military terms. Current events suggest that ethnic conflict, environmental degradation, pollution, and famine leading to civil unrest or massive migrations of refugees, will constitute threats to both social stability and the preservation of the planet's biodiversity. Stopping deforestation or augmenting food production capabilities in deficit areas should directly and substantially contribute to the security of society. It also should help prevent armed conflict. Activities that may reduce the likelihood of conflicts leading to war include allocating international resources to environmental monitoring and impact assessment, protection of economically important species, quick response to disasters and accidents, energy conservation, and the minimization and management of waste. As Thacher (1984) put it, "Trees now or tanks later."

The possibility of using protected areas along their borders as ways of promoting peace is now recognized by at least some countries (e.g., Hanks, 1998). In many countries, boundaries are found in mountainous areas that also tend to be biologically rich. They often encompass a great variety of habitats and ecosystems within relatively small areas, consequence of differences in elevation, microclimate, and geological factors. While such ecologically diverse areas are often particularly important for conservation of biodiversity, they also are frequently sanctuaries in war, especially civil wars and guerrilla wars.

National frontiers are especially sensitive areas where conflict is endemic and biological resources often are especially rich. The idea of establishing protected areas on both sides of borders of countries that have tense relations—as so-called "peace parks"—has attracted considerable attention. Peace parks provide a symbol of the desire of the bordering countries to deal with many of their problems in a peaceful way (see, for example, Thorsell, 1990; Westing, 1993, 1998). Zbicz and Greene (1998) have found that transboundary protected areas cover well over 1.1 million km^2, representing nearly 10% of the total area protected in the world (see Table 10.3). This also demonstrates how much of the world's land area devoted to biodiversity conservation is in remote frontier areas where risks of war historically are highest.

Such peace parks are far more than a fond hope. Peru and Ecuador fought three territorial wars in the 20th century, but then Peruvian President Alberto Fujimori and Ecuadorian President Jamil Mahuad developed an innovative plan to resolve their violent border dispute in 1998. This plan included creation of two national peace parks near the most contested stretch of their frontier. Four mediators, the United States, Argentina, Brazil, and Chile, helped resolve the hottest regional dispute in South America through binding arbitration. Ecuador was granted free trade and navigational access to Amazon River through the Napo River that starts in Ecuador and ends in Peru. Ecuador achieved many of its economic goals, although not sovereignty over the disputed territory.

Brock (1991) concludes that although peace parks have probably had relatively little independent effect on international relations, transfrontier cooperation on biodiversity

Table 10.3. *Trans-frontier protected areas in tropical forest regions.*

Continent	*Transfrontier protected area complexes*	*Designated protected areas*
Africa	39	110
Asia	31	74
Latin America	35	89
Total	105	273

issues has the potential to develop into an important factor in at least regional politics by helping to internalize norms, establish regional identities and interests, operationalize routine international communication, and reduce the likelihood of the use of force.

Peace parks, however, will need special facilities to be able to be effective. The professional development of national staff and strong site-based conservation programs are the best preparation for conservation in the face of regional instability (Hart & Hart, 1997). Key to the initiative, however, is that these peace parks must be tied to an international structure that endures when nations crumble. Establishing a fund that provides for continued professional development and support for field activities by the staff of protected areas during crisis periods will have to be part of that. Such support might be focused on specific sites of international biological significance with the goal of developing semiautonomous management within those areas. The mission of the proposed fund would be to build professional identity in national staff where national institutions have failed and to facilitate their reintegration into conservation activities after the crisis has passed.

As a concluding remark, the relationship between war, conflict, and biodiversity is highly complex. Individuals make multiple, mutually constraining decisions that are shaped by interacting environmental and social conditions, all of which themselves have multiple interrelationships. People often learn through conflict when fundamental interests are challenged. As Lee (1993) points out, "Conflict is necessary to detect error and to force corrections. But unbounded conflict destroys the long-term cooperation that is essential to sustainability. Finding a workable degree of bounded conflict is possible only in societies open enough to have political competition." Or, to phrase it differently, the solution to war can also help conserve the natural resources upon which people depend.

REFERENCES

Abdulla, R. (1997). Protected areas during and after conflict. Nimule National Park: A case study. In IUCN (Ed.), *Parks for Peace Conference Proceedings* (pp. 195–199). Gland, Switzerland: IUCN.

Brock, L. (1991). Peace through parks: The environment on the peace research agenda. *Journal of Peace Research, 28*(4), 407–423.

Crosby, A. (1986). *Ecological Imperialism*. New York: Cambridge University Press.

Dillon, T. C., & Wikramanayake, E. D. (1997). Parks, peace and progress: A forum for transboundary conservation in Indo-China. *PARKS, 7*(3), 36–51.

Edgerton, R. B. (1992). *Sick Societies: Challenging the Myth of Primitive Harmony*. New York: The Free Press.

Ember, C. R., & Ember, M. (1992). Resource unpredictability, mistrust, and war. *Journal of Conflict Resolution, 36*(2), 242–262.

Fairhead, J., & Leach, M. (1995). False forest history, complicit social analysis: Rethinking some West African environmental narratives. *World Development, 23*(6), 1023–1035.

Ferguson, R. B. (1989). Ecological consequences of Amazonian warfare. *Ethnology, 28*, 249–264.

Flannery, T. (1994). *The Future Eaters: An Ecological History of the Australasian Lands and People.* New York: George Braziller.

Hanks, J. (1998). Protected areas during and after conflict: The objectives and activities of the peace parks foundation. *PARKS, 7*(3), 11–24.

Harbinson, R. (1992). Burma's forests fall victim to war. *The Ecologist, 22*(2), 72–73.

Hart, T. B., & Hart, J. A. (1997). Zaire: New models for an emerging state. *Conservation Biology, 11*(2), 308–309.

Heider, K. (1970). *The Dugum Dani: A Papuan Culture in the Highlands of West New Guinea.* Chicago: Aldine.

Hill, J. D. (1994). Alienated targets: Military discourse and the disempowerment of indigenous Amazonian peoples in Venezuela. *Identities, 1*(1), 7–34.

Homer-Dixon, T. F. (1994). Environmental scarcities and violent conflict: Evidence from cases. *International Security, 19*(1), 5–40.

Kane, H. (1995). The hour of departure: Forces that create refugees and migrants. *WorldWatch Paper, 125*, 1–56.

Keeley, L. H. (1996). *War Before Civilization.* New York: Oxford University Press.

Knauft, B. M. (1990). Melanesian warfare: A theoretical history. *Oceania, 60*, 250–311.

Lee, K. N. (1993). *Compass and Gyroscope: Integrating Science and Politics for the Environment.* Washington, DC: Island Press.

Martin, P. S., & Klein, R. G. (Eds.) (1984). *Quaternary Extinctions: A Prehistoric Revolution.* Tucson: University of Arizona Press.

McClure, H. E. (1974). *Migration and Survival of the Birds of Asia.* Bangkok: US Army Component, SEATO Medical Research Laboratory.

McNeely, J. A. (1994). Lessons from the past: Forests and biodiversity. *Biodiversity and Conservation, 3*, 3–20.

McNeely, J. A., & Wachtel, P. S. (1988). *Soul of the Tiger: Searching for Nature's Answers in Southeast Asia.* Singapore: Oxford University Press.

Meggitt, M. (1977). *Blood is their Argument: Warfare among the Mae Enga Tribesmen of the New Guinea Highlands.* Palo Alto, CA: Mayfield.

Myers, N. (1979, Dec 8). Wildlife and the dogs of war. *The Daily Telegraph*, London.

Nietschmann, B. (1990). Battlefields of ashes and mud. *Natural History, 11*, 35–37.

Pearce, F. (1994, Dec 3). Soldiers lay waste to Africa's oldest park. *New Scientist* p. 4.

Ponting, C. (1992). *A Green History of the World: The Environment and the Collapse of Great Civilizations.* New York: St. Martin's Press.

Rappaport, R. A. (1984). *Pigs for the Ancestors. Ritual in the Ecology of a New Guinea People.* 2nd. ed., enlarged. New Haven, CT: Yale University Press.

Strada, G. (1996). The horror of land mines. *Scientific American, 274*(5), 26–31.

Thacher, P. (1984). Peril and opportunity: What it takes to make our choice. In J. A. McNeely & K. R. Miller (Eds.), *National Parks, Conservation, and Development: The Role of Protected Areas in Sustaining Society* (pp. 12–14). Washington, DC: Smithsonian Institution Press.

Thorsell, J. (Ed.) (1990). *Parks on the Borderline: Experience in Transfrontier Conservation.* Gland, Switzerland: IUCN.

Westing, A. H. (1976). *Ecological Consequences of the Second Indo-China War.* Stockholm: Almqvist and Wiksell.

Westing, A. H. (1982). The environmental aftermath of warfare in Vietnam. In *World Armaments and Disarmament: SIPRI Year Book 1982*, SIPRI (pp. 363–389). London: Taylor and Francis Ltd.

Westing, A. H. (1993). Transfrontier reserve for peace and nature on the Korean peninsula. In IUCN (Ed.), *Parks for Peace Conference Proceedings* (pp. 235–242). Gland, Switzerland: IUCN.

Westing, A. H. (1998). Establishment and management of transfrontier reserves for conflict prevention and confidence building. *Environmental Conservation*, *25*(2), 91–94.

Zbicz, D. C., & Greene, M. (1998). Status of the world's transfrontier protected areas. *PARKS*, *7*(3), 5–10.

INDEX

2,3,7,8-tetrachlorodibenzo-para-dioxin; TCDD 155
2,4,5-trichlorophenoxyacetic acid; 2;4;5-T; 2.4.5-TCP 154,155
2,4-trichlorophenoxyacetic acid; 2,4-D 155
abandoned habitats 134
access roads 104
access to land 41, 70, 85, 89
accountability 8, 33, 111, 119, 122
Action Plan for Forest Law Enforcement, Law and Governance 118
aerial herbicide 142
afforestation 22
Afghanistan 2, 15, 141, 159
Africa Action Plan 118
Africa, postcolonial 53
African civil wars 52, 123
African conflicts 96, 118
African Forest Law Enforcement and Governance 105
African forest, Africa's tropical forests 1, 94, 97, 105, 112
African leaders 105, 109
African state forms 94, 109, 112
African Union 52, 95
African Union peacekeepers 94
African Wildlife Foundation 108
Africa's First World War 95
Agent Blue 155
Agent CS 152
Agent Orange 153, 155, 156, 157, 158
Agent White 155
agricultural collectives 85
agricultural production 85, 88, 133, 168
aid conditionality 33
Amazonian forest 141
American advisers 78
American Ambassador Nolting 151
American bison 166
American soldiers 10, 149, 152, 156
Americo-Liberian community 43
ammonia 136
ammonium thiocyanate 150
amnesty 67
Anastasio Somoza 62
Andean forests 139, 141
Andes; Andean region 2, 133, 137, 139, 141, 142, 145
Andre Castro United Front; FUAC 64, 67, 68, 69, 71
Angola 37, 56, 94, 95, 98, 99, 103, 113
anthrax 150
antidrug policies 142
antiforeigner attacks; antistranger rhetoric; antiforeigner rhetoric 46, 55
antiguerrilla weaponry 156
appropriation of resource 7
aquatic and terrestrial habitats; aquatic ecosystems 137, 143
aquifers 167, 168
armament 4, 42, 49
armed conflict 2, 5, 9, 10, 12–15, 35, 37, 73, 74, 94, 114, 117, 118, 120, 126, 128, 12–134, 139,140, 145, 159, 169
armed groups 17, 23, 24, 66, 95, 117–120, 122–134, 127, 128, 134, 140
arms proliferation 118
arms sales; arms trade 108, 122, 126
Arnoldo Aleman 77
Asian Development Bank 162
Asian forest 1
asset freezing 108
Association of Southeast Asian Nations; ASEAN 20, 127
Avança Brasil 59, 73

B-29, bombers 150
backward fuel switching 102
bacteriological weapons 150
bamboo forest 161
ban on log exports; ban on timber exports 25, 128, 129
banditry; bandits 12, 52, 79
Bangladesh 161
barking deer 163
barren forest 87
battle areas 161
belligerents; belligerent groups 9, 13, 18, 94, 96, 103, 108, 133
Belom Forest Reserve 164
bilateral accords 127
bilateral and multilateral approaches 108

Biodiversity Support Programme 108
biodiversity; biological diversity 2, 10, 65, 71, 72, 77, 93, 94, 103, 107, 124, 134, 136, 141, 142, 159–170
bio-hotspot 96
biological diversity, global 141
biotic degradation 140
birth defects 156
blood diamonds; conflict diamonds 6, 13, 52, 58, 61, 96, 101, 107, 126, 131
Bolivia 2, 15, 61
bombing; bombs 17, 20, 78, 130, 149, 150, 157, 163, 164
border control 52
Bosawas; Bosawas Biosphere Reserve; Bosawas National Reserve 57–59, 62–72
Botswana 95, 114, 115
botulin 150
boycott 13, 38, 52
Brazil 2, 61, 62, 73, 167, 169
Brazilian Amazon 58, 73
British and North American culture 62
British commandos 52
British Special Forces 42
Bulgaria 96
Burkina Faso 39, 42, 96, 97
Burma 61, 117
Burman people 127
Burma's forests 171
Burundi 2, 17, 94, 95, 98, 99, 102, 162
bush camps 42, 43
bush knowledge 51
bushmeat 93, 103, 123, 133

Ca Mau peninsula; Cape Ca Mau 153–154
cacodylic acid 155
Cambodia 22–26, 28, 30–36, 117, 120, 122, 125, 129–131, 152, 159, 162, 163
Cambodia Study Group 25
Cambodian companies 23
Cambodian -Thai border 130
Cambodia's National Assembly 29
Cambodia's People's Party, CPP 21, 22, 25–29
Cambodia's Royal Government; RCG 25, 28, 30
Cameroon 98, 99, 101, 102, 104, 105, 107, 108, 112, 114
Cameroon Environmental Protection Support project 107
Canada 67, 73, 78, 97, 104, 107
capital accumulation 18, 22
capital density 60
capital investment 128
capitalism 24, 27
capitalism, crony 29
capitalism, global 111, 112
carbon store 93, 113
CARE 161
Caribbean forests 141
Caribbean region 133
cartels, criminal; drug cartels 61, 96, 159
cash crop; cash crop production, cash crop producer 84, 102, 110, 112, 134
Catholic 63
cattle ranching 133, 139, 142
ceasefire agreement 94
center-periphery style state forms 110
Central Africa 14, 37, 93–95, 97–105, 11–114, 161
Central African conflict 98
Central African deforestation 115
Central African landscape 104
Central African rainforest 103
Central African Regional Program for the Environment; CARPE 103
Central African Republic 2, 13, 95, 98, 99
Central and West Africa 38, 99, 102
Central and West African conflicts 96
central state 110, 112
certification regimes; forest certification schemes 104, 125–127
Chad 94
chain of custody 108, 117, 126, 127
charcoal 3, 4, 102
Charles Taylor 23, 39–42, 45–48, 93, 96, 97, 99–101, 110, 114, 171
chemical warfare 150, 151, 156
chemical waste 137, 140
Cheyenne 166
chiefs 41, 43, 45, 46
China 15, 21, 23, 79, 100, 161, 162, 163
chloracne 157
Choco 137, 141, 142
Christianity 43, 44
CIFOR 103, 117
civil conflict 47, 53, 160
civil servants 27, 44
civil war 1, 8, 9, 14, 21, 22, 37–43, 45, 47–52, 58, 62, 70, 72, 76, 78, 94–96, 110, 113, 120, 124, 159, 161, 162
civilian vigilante groups 44
clear felling 161
climate change 103, 168
coca 10, 134–136, 138, 140, 141, 144, 145, 147
coca base, cocaine 134, 136, 144–147
cocoa 46

coercion 50
coffee 46
Cold War 19, 33, 37, 49, 58, 66, 122
Cold War, post- 47, 97, 109, 110
collared peccaries 164
collective action 93, 160, 168
collective decision making 44
collective tenure 41
Colombia 1, 2, 9, 61, 133–143
colonial rule 46
colonialism 110, 131
colonists 60, 166
colonization 61, 70, 134, 142, 143
coltan 96, 110, 113
combatants 2, 7, 10, 11, 39, 41, 45, 53, 56, 70, 75, 120, 121, 163
combatants, ex- 41, 53, 64, 100, 101, 114, 124
commercial agriculture 60
commercial and military alliances 120
commercial farmers 61
commodification of forest 22, 30
commodification of nature 18
commodity networks 44
common property resource management 103
communal lands 122
communications technology 4
communities 1, 4, 41, 45, 46, 49, 53, 69, 70, 88, 90, 102, 108, 112, 161, 165, 166
communities, forest-; communities, forest-based 11, 93
communities, indigenous; ethnic communities 45, 46, 69, 70
communities, local 28, 33, 49, 102, 107, 122–124, 134
community forestry 112
concession (fishing) 69
concession (forest, logging, timber) 10, 13, 24, 27, 29, 31, 68, 69, 101, 105, 118, 119, 121, 122, 163
concessionnaires (forest) 27, 29, 105, 119
conditional disbursement 129
conflict commodities 118, 120, 125, 128, 130
conflict resolution 1, 33, 58, 72, 131, 170
conflict timber 2, 58, 61, 96, 112, 113, 117–131
congenital abnormalities 155
Congo Basin Forest Partnership; Congo Basin Forestry Program 104, 108
Congo river; Congo basin 1, 103, 106, 108, 132
Congress of the Armies of the Americas 167
conscription 79
conservation forest 87
conservation function 66
Conservation International 108, 112, 145
consumer countries 118, 127, 130
Convention on Biological Diversity; CBD 159, 160
Convention on International Trade in Endangered Species; CITES 125
Convention, Environmental Modification 5
Convention, Geneva 5
cooperation among states 93
corporations; transnational corporations 96, 129, 131
corrupt authorities 24
corruption 10–12, 12, 19, 22, 27, 29, 31, 34, 45, 111, 118, 119, 127
corvee 79
Côte d'Ivoire, Ivory Coast 39, 44–46, 48, 50, 52, 54–56, 94, 96, 97, 100, 101, 110
coup d'etat 20, 61
creoles 43
crocodiles 164
crops, annual 133
crops, area-extensive 142
cross-border forest regions 48
cross-border trading 48
customary authority 40
customary rule 18, 87

Danish Forestry Company; DLH group 100
debt forgiveness 112
debt repayment 102
debt/environment swaps 112
decommissioned officers; decomissioned soldiers 64, 69, 70
defoliants; defoliation 12, 137, 149–158
deforestation 139–145, 147, 160, 165, 168
degradation of infrastructure 96, 122
demobilization 53, 70, 125
demobilized fighters 65, 69
Democratic Republic of Congo; DRC 2, 52–55, 94–108, 113, 115, 117–131, 161, 162
democratization 32, 112, 144
Denmark 100
deregulation 111
desertification 160
destruction of wildlife habitat 123
development agencies 18, 20–22
development assistance 66, 71, 125
development initiatives 122
development poles 66
diamonds 7, 9, 12, 14 37, 40, 42–44, 47, 52, 53, 55, 56, 97, 100, 101, 109, 110, 118, 121, 126, 127, 132, 154
diarrhea 94

Dinoxol 155
dioxin 150, 153, 155–158
disarmament 40, 42, 49, 67, 68, 125
displaced, displace populations; displacement of civilian populations 46, 78, 80–82, 88, 94, 95, 103, 133
dissidents, dissident groups 7, 39
donor pressure 130
donors; international donors 20, 26, 29, 33, 103, 108, 110–113, 127, 129, 130
Dow Chemical 154
DRC's 2002 Forest Code 106
drug trafficking 132, 142, 145
dry tropical forests 98
Dutch disease 47

ecocide 161
ecological attacks 166
ecological degradation, ecological impacts 61, 88
ecological marginalisation 38, 113
economic agendas 37, 38, 49, 51
Economic Community of West African States Cease-fire Monitoring Group; ECOMOG 9, 40, 42, 52, 97, 127
Economic Community of West African States; ECOWAS 127
economic decline 44, 110
economic development 46, 90, 94, 95, 103, 104, 123, 134, 145
economic globalization 111
economic opportunism 8, 49
economic pathologies 94
economic reconstruction 18
economic sanctions 24, 25, 33
ecosystem 1, 10, 102–104, 133, 134, 136, 137, 139, 141, 143, 144, 157, 166, 169
ecosystem function 133, 141
ecosystem stability 104
edible nuts 149
effluents; effluvia 135, 136, 140
elites 24, 42, 55, 61, 112
elites, central 33
elites, party 28
elites, political 20
elites, national 71
elites, bureaucratic 33
elites, politico-military 31, 32
elites, powerful 38
elites, ruling 32, 41
Ellen Johnson-Sirleaf; Johnson-Sirleaf 97, 111
embourgeoisement 28, 33
emigrants; immigrants 7, 44, 46, 50, 83
empty forest phenomenon 103
endangered animal and plant species; endangered tree and plant species 65, 87, 123, 125, 142, 161
endemism; endemic plants 103, 133, 141
environmental authorities 134
environmental conservation 1, 13, 73, 147, 172
environmental crisis 20
environmental damage 5, 142, 143, 160
environmental decline; environmental degradation 7, 22, 54, 57–59, 61, 93, 109, 143, 167, 177
environmental destruction 2, 5, 7, 58
environmental impact 3, 13, 15, 123, 132, 140, 143, 144, 150
environmental imperialism 103
environmental protection 134, 143
environmental quality 133, 141
environmental scarcity 6, 11, 14, 38, 55, 113, 171
environmental security 6, 8, 38, 113, 114
environmental stress 168
eradication 135–137, 140–144
Eritrea 94
Ethiopia 94, 150
ethnic animosities 50
ethnic categorizations; ethnic differentiation 40, 51
ethnic cleansing 46
ethnic divisions 50
ethnic fractionalization 50, 51
ethnic grievance 50
ethnic group 4, 12, 41, 43, 46, 50, 79, 82–85, 89, 90
ethnic resentment 50
ethnic scapegoating 50, 53
ethnic vocabularies 51
ethnically inspired ideologies 37
ethno-nationalism 19
excess mortality 94, 95
Executive Outcomes 42, 97
expropriation of lands 123
extinction risk; extinction 103, 142, 171
extirpation 133, 142
extortion 7
extra budgetary activities 119
extrabudgetary income 100
extra-contract payments 121
extreme conflict—tropical forest link 6–9, 12, 17, 31, 109, 118, 126

failed state 33, 55, 121
FAO 104, 105, 108
farm cooperatives 63
firewood 102, 161, 164, 167
Foday Sankoh 42, 44
food insecurity 22

forced displacement 134
Forces Nouvelles 45, 96, 110
foreign companies; forest enterprises 24, 31, 79, 100, 105, 121
foreign investors 27
foreign militaries 101
forest administration 108
forest antelope 137
forest areas of Colombia 1
forest categories; land- and forest use classification 79, 87
forest commodification 20, 41
forest concessionaires 29, 119
forest conservation 73, 94, 106
forest consultative councils 107
Forest Crime Monitoring Unit 29
forest department 23, 27
forest dwellers 20
forest frontier 58, 74, 139
forest land distribution 11
Forest Law Enforcement and Governance; Forest Law Enforcement, Law and Governance 105, 118
Forest Law; forestry law; forestry legislation 28, 87, 105, 118, 119
forest loss 78, 88, 101, 111
forest management 4, 10, 75, 78, 79, 86–90, 93, 106, 107, 112, 123, 125, 131, 146
forest management, postwar 78, 79
Forest Principles 108
forest product collection 89
forest regions in Bolivia 2
forest regions of Burundi 2
forest resource boundaries 77
Forest Stewardship Council 126
forest survey Laos 79, 87
forest wars 38, 47, 109
forest-dwelling people 167
forestry industry 100, 104, 107
forestry rent 31
forestry sector 22, 26, 29, 31, 32, 121, 124, 129
forests in Lao PDR 75
fragmented forests 75
France 74, 100, 112, 131, 132
freezing bank accounts 52
French colonial armies 164
French intervention 80
French protectorate 19, 21
French troops 45
frontier economies 59
frontier forest 4, 62, 99, 103, 109
Frontier Forests Initiative 99, 109
frontier formation 60, 61
frontier, agricultural 73, 82, 133
frontier, Amazonian 62
frontier, corporate 60
frontier, populist 60
frontiers, modern 58
fuelwood consumption 97
FUNCINPEC 21, 25–29

Gabon 98, 99, 104
Gabon basin 103
gasoline 135, 136
GDP 20, 21, 40, 133
gems 24, 25
General Chavalit Yongchaiyut 25
Geneva Accords 151
genocide 21, 94, 102, 126
Germany 100
Ghadaffi 39, 43
Ghana 39, 43, 98, 99, 102, 112
global conflicts 109
global public goods 103
Global Witness 28, 29, 30, 107
globalization 108, 111, 113, 143
glyphosate 136, 142, 143, 146
GNP per capita growth 104
Gola forest reserve 48
gold 37, 40, 45, 61, 97, 121
Golden Triangle 61
governance interventions 94
governance reforms 122
governance, bad 48,109
governance, forest 10, 12, 13, 117, 129
governance, formal 19, 31
governance, frontier 19
governance, global 108, 111, 113
governance, good 62, 95
governance, poor 6, 7, 12, 119, 127, 130
governance, postwar 12
governance, resource 123
governance, Sandinista 64
governance, shadow 19
government authorities 121, 126
government control 10, 12, 42, 62, 86, 89
government institutions 11, 53, 60, 65
Government of Cambodia; Royal Government of Cambodia 25, 26, 28, 31, 35
Government of Laos, Royal; Royal Lao Government 78, 81, 85, 89
Government of Liberia, National Transition 97
Government of the DRC 95, 105
government soldiers 44

government, central 4, 5, 12, 65, 69, 70, 71, 104
government, centralized 61
government, emerging 60, 70
government, external 60, 61
government, imposed 60, 70
government, local 61, 72
government, national 46, 53, 62, 65, 66, 68, 70–72
government, postwar 75
government, socialist 78
Government, Venezuelan 167
governments, African 109, 118
Great Lakes Region of Central Africa 95
Greece 100
greed 8, 37–43, 45–49, 51–53
Green book 43, 44
green reconstruction 32
Greenpeace 103, 111, 113, 114
grievance 7, 8, 11, 14, 37–39, 41–51, 53
grileiros; land grabbers 61
GTZ 65, 71, 79
guerilla 8, 10
Guinea 40, 41, 46, 96, 98, 99, 103, 104, 117, 164–167
Gurkha Security Guards 97
Gus van Kouwenhoven 13
Gyude Bryant 40

habitat destruction 160, 165
herbicide 5, 10, 136, 142, 144, 149, 152, 154, 155, 157
heroin 135, 141, 145
high-value products 4
high-value tropical timber; high-value timber 119 120
hit-and-run attacks 42
HIV/AIDS 56
Hmong 80, 84
Ho Chi Minh Trail 154, 163
Hodgkin's disease 157
Home Depot 94
Honduras 57, 59, 62–64
Horn of Africa 37
hostile groups 75
Human Development Index 104
human induced fragmentation 106
Human Rights Watch 30
human rights, human rights abuses 17, 22, 167
humanitarian disaster 73, 95, 108, 131
Humberto Ortega 67
Hun Sen 21, 27, 29, 30
hunter-gatherers 4
Hutu rebels 17
hyperliberal world order 94

Ibraham Kamara 42, 43
ideology 42, 43, 63, 160
iguanas 164
IKEA 104
illegal logging 11, 27, 29, 30, 32, 34, 35, 55, 99, 108, 114, 118
illicit crop; illicit drug crop 2, 10, 133–145
illicit drug; prohibited drugs 127, 13, 133–135, 140, 142, 143, 146
IMF 20, 35, 129
importing countries 33, 130
impoverished peasants 29
impoverishment of upland farm households 87
incense 4
India 2, 5, 14, 161, 162
indigenous military commanders 69
indigenous people; indigenous populations 11, 40, 41 62, 69, 74, 107, 165, 167
indigenous territorial and civil rights 69
individual title deeds 40
Indochina War; Indochina conflict 10, 21, 78, 79–89, 162, 164
industrial round wood 102
inequitable land distribution 102
informal market networks 123
infrastructure destruction 96
institutional vacuum 58
instrumentalization of war 19
insurgency, counter 53
insurgents; insurgent groups 25, 28, 37, 36–39, 47, 49, 53, 65, 66, 118, 120, 122, 125, 129, 161, 164
integrated natural resource management 108
intellectuals, urban 43, 53
intercommunal fighting 46
interethnic tension 43, 127
intergenerational exploitation 49
intergovernmental bodies 52, 118
intergovernmental institutions 93
intergovernmental missions 52
internally displaced; internal displacement 95, 96, 103
international agencies 1, 32
international arrest warrants 52
International Commission on Peace and Food 159
international community 5, 6, 13, 18, 22, 24, 31–33, 35, 40, 52, 53, 65, 72, 108, 126, 130
International Conference on the Great Lakes Region 95
international conflict 120

International Criminal Court 126
international development agenda 33
international development community 57
international intervention 1, 13
international measures 52, 118, 143
International Monetary Fund; IMF 20, 35, 129
international trade 4, 20, 25, 32, 99, 118, 125, 126, 130
international treaties 125
interstate cooperation 109
Inter-state Forestry Ministers' Working Group 104
intervillage differences 77
invading armies 3
invasion 160
Iran-Contra affair 63
Iraq 1, 159
irrigated rice production 82
irrigation channels 163
Islam 43
Italy 100, 112
Ivorian army 48
ivory 103

Joint Chiefs of Staff 150, 155, 158

Karen National Union; Karen 17, 161
Kenya 95, 111
Khmer People's National Liberation Front 21
Khmer Rouge 17, 18, 20–28, 31–35, 120, 129
Kimberly process 51
kleptocracy 111

land availability 58, 139
land conflicts 61, 69
land consolidation 134
land distribution 11, 71, 102
land hunger 102
land tenure 46, 58, 60, 61, 64, 70, 140
land titling scheme 70
landslides 140
land-use classification 86, 89
Lao PDR; Laos 2, 9, 10, 14, 61, 75–89, 152, 154, 156, 161, 163
large companies 60
Latin American forest 1
laundering funds 121
Laurent Kabila 101
law enforcement 24, 29, 52, 58, 61, 65, 66, 67, 105, 118, 119, 140
lawlessness 49, 58, 61, 66, 161
legal dualism; legal duality 40
legal origin 126
legal pollution 141
legal verification 126
Lesotho 110
liberal war 114
liberalization, market; economic liberalization; liberalization and deregulation 31, 84, 85, 111, 112, 114, 145
Liberia 2, 13, 37–43, 45, 46, 48–50, 52, 54, 93, 96–101, 104, 108, 110, 111, 113, 117, 118, 122, 125, 127–129, 162
Liberia, Greater 99, 100
Liberian Forestry Development Authority; LFDA 100
Liberian Forestry Initiative 104
Liberian timber 108
Liberians United for Reconciliation and Democracy; LURD 39–41, 122
Liberia's American-Africans 40
Liberia's timber industry 40
Libya 39, 42, 96, 97, 110
liquid waste 136
lobbyists 108
local and global imaginaries 106
local livelihoods 123
local merchants 60
local strongmen 43
log extraction, log production 4, 78
logging 10, 11, 13, 17–20, 22–33, 45, 55, 60, 69, 78, 79, 81, 82, 87–89, 93, 97, 99–101, 104, 105, 107–109, 114, 118, 120–125, 127, 130, 131, 161–164
logging ban 23
logging companies 17, 24, 93, 101, 104, 105, 109, 121, 122, 127, 163
logging quotas 27, 79
logging revenues 28–30
logging sector 17, 22, 27, 32
Lon Nol 31
lootable resources 47, 49
looting 42, 44, 48, 50
looting of natural resources 120–122
low intensity conflicts; low level conflicts 94, 119, 100
lung cancer 157
Lusaka accord 95

malaria 94
malathion 152
Malawi 102, 111
Mali 41
malnutrition 94
Mandingo people; Mandingo 41

mangrove 98, 103, 153, 154, 161
mantled howler monkey 164
market access 120, 124
market economy 79
Mayagna Indians 63
Medecins Sans Frontiers 109, 161
medievalism of African politics 100
Mekong River 21, 78, 80, 81, 83, 84
Mekong River Commision 79
mercenaries; mercenary groups 42, 55, 101
mestizo 62, 70
migrants; migration; migrant settlers 21, 45, 46, 76, 80–86, 89, 139, 140, 171
Migratory Animals Pathological Survey 163
migratory birds 154, 163
militarized society 50
military camps 50
military hardware 4
military interests 121
military intervention 26, 38, 52, 121
military observers 26
militia, demobilized 11
militia, patriotic 45
militia, roaming 11
militias, armed 122, 128
militias, company 122, 123
militias, Interhamwe 125
militias, Mai Mai 125
militias, private 120
mining 24, 25, 40, 45, 60, 65, 69, 96, 97, 101, 105, 133, 139, 145
Ministry of Agriculture and Rural Development of Vietnam 154
Miskito Indians 63
MISURA; MISURISATA 62–66, 68, 69
mode of accumulation 37
monitoring technologies 125
Mon-Khmer ethnic group 75
Monsanto 154
moratorium on the export of logs 24
Moravian church 63
mountain gorilla 107, 161
Mouvement de Liberation Congolese 95
Movement for Democracy in Liberia; MODEL 39, 40, 45, 49
Mozambique 2, 13, 37, 61, 98, 102, 124
multilateral international agreements 127
multiparty elections 111
multiple myeloma 157
Myanmar 2, 117, 127, 128, 159, 161
Myanmar's forest 128
myelogenous leukemia 157
mystic space 94, 103, 106
Nam Ngum reservoir 78
Namibia 95
napalm 153
narcotics 135, 144
nation building 8, 53
National Geographic Society 103, 106, 109
National Institute of Colonization and Agrarian Reform (Brazil) 61
national park 10, 29, 49, 72, 123, 161, 162, 164, 170, 171
national patrimony 12
National Patriotic Front; NPFL 39, 40, 42, 45–49, 127
native languages 43
natural disaster 133, 140, 165
natural resource control 119
natural resource exploitation 4, 18, 22
natural resource hot spots 48
natural resource revenues 53, 118
natural resource sectors 40, 48, 49, 53, 119
nature protection groups 18
neoliberal 8, 19, 32, 109, 111, 113
neoliberal globalization 111, 113
neoliberal prescriptions 32
neoliberal reforms 32
neoliberal world order 109, 111
neo-patrimonial networks 53
Netherlands 13
New Guinea 165–167, 171
Nicaragua 2, 9, 57–72, 160, 162–164
Nicaraguan army 58, 67, 69
Nicaraguan Democratic Force; FDN; Contras; ex-Contras 57, 58, 62–67, 71, 73, 106, 107, 127, 129, 162
Nigeria 2, 96, 113
Nigerian peace keeping forces 44
night monkey 164
Nimba County 39, 41
Nimba mountains 96
Nixon administration 156
non-Hodgkin's lymphoma 157
nontimber forest products; nontimber species 24, 102, 104
Norodom Ranariddh 26, 27
North Atlantic Autonomous Region 62, 64
North Vietnamese forces 80, 82
Northern Africa 98
Northern Front 3-80; FN 3-80 64, 66–69, 71

oil palm 162
open acces resources 75

Operation Ranch Hand 149–151, 153, 154, 156–158
organic chlorine herbicide 154, 155
organic compounds 140, 146

paddy fields 82, 85
paramilitary groups; paramilitaries 119, 133, 134, 144
Paris Peace Agreement 20
park rangers; park staff 69, 162
Pathet Lao 78, 81, 82, 163
patrimonialism; patrimonial politics 41, 48
patronage of armed movements 122
peace agreement 20, 21, 24, 40, 44, 64, 68, 125
peace building 22, 53
Peace Corps 108
peace park 10, 13, 159–170
peace process 24, 31, 96, 107
peace settlement 71, 128
peacekeepers 13, 94, 95
Peninsular Malaysia 164
perennial crops 133
Peru 2, 61, 141, 169
petroleum; oil 12, 25, 96, 104, 113, 120, 153, 160
physical maiming 96
picloram 155
pit-sawing 100
Plan Colombia 141, 145
plankton 155
plantation 97, 102, 110, 144, 162
Pol Pot 23, 35
policing forest areas 122
policy interventions 107
political ecology; political ecologist 6, 7
political economy 19, 24, 28, 33–35, 54–56, 108, 127
political settlement of conflict 124
pollutants 140
pollution of watercourses 140
Polynesia 165
population growth 38, 97, 168
population movement 76, 79, 80, 81
population pressure 83, 93, 167
porphyria cutanea tarda 157
Portugal 100
postcolonial era 4, 19
postconflict 10, 11, 13, 94, 124, 127, 129, 162
postconflict depredation 13
postconflict reconstruction 129
postconflict scenarios 127
poverty 7, 15, 12, 96, 111, 167, 168
powerless group 38
precious metals 37
predation of forest resources 11
preindustrial societies; prestate societies 165, 167
President Kennedy 150
primary commodity exports 47
Prime Minister Chavalit 31
Prince Johnson 39, 40
pristine forest narrative 106
pristine wilderness 107
private leasehold 102
privatization 111
production forest 87, 107
property rights; de facto property rights; de jure property rights 12, 41, 59–61, 72, 73
prostate cancer 157
protected area system 77, 104
protection fee 23, 28
protection forest 87
proxy conflicts 37
Pseudonovibos spiralis 137
psychological warfare 152
Putumayo River 141

Rainforest Action Network 103
Rand Corporation 155
rapprochement 23
Rassemblement Congolais pour la Democratie 95
Rassemblement des Republicains 39, 44
Reagan administration 63
realpolitik 32
rebel armies 120, 122
rebel leaders 47, 50, 52
rebel narratives 38
rebellion 7, 8, 37, 42, 43, 47–50, 52, 108, 110, 127
red backed squirrel monkey 164
Red Cross 161
refugee camps 46
refugee movements 81
refugees 2, 9, 11–13, 15, 23, 41, 46, 67, 76, 78, 80–82, 88, 95, 96, 102, 103, 107, 132, 149, 161, 165, 168, 171
regeneration forest 87
regional organizations 127
regulatory agencies and judiciaries 119
reintegration 53, 71, 170
rentier state 47, 110
rents 1, 46, 96, 105
repatriation 49

repression 143, 160
Republic of Congo; Congo Brazzaville; Congo 2, 94, 95, 98, 124, 131, 159
resource abundance 38, 47, 101
resource access 68, 77, 88, 89
resource appropriation 7, 38
resource boundary 87
resource conservation 111
resource consumption 168
resource control 7, 31, 119
resource curse 101, 109
resource depletion 7
resource distribution 7, 38, 130
resource raiding; resource plunder 100, 101
resource scarcities 9, 38, 89, 160, 168
resource rich regions 40, 49, 168
respiratory infection 94
Revolutionary United Front; RUF 39, 42–44, 47–50, 93, 100
rhino horn 103
Rhizophoraceae; Rhiziphora apiculata 153, 154
rice cultivation 82
rice paddies 150, 154
rich natural resources 58
Rio San Juan 9, 62
road construction 58, 62, 70
road networks 106
Robert Mugabe 101
Romania 96
Royal Cambodian Armed Forces 120
Royal Timber Company 100
rubber business 100
rule of law 9, 27, 32, 53, 108, 119, 123, 124, 126, 162
rural leaders 58
Russia 96, 97
Rwanda 2, 13, 15, 73, 94, 95, 98, 99, 101, 102, 121, 132, 161
Rwandan genocide 94
Rwandan refugees 13, 15, 132

Sahel 98
Saigon newspaper 151, 156
Samual Doe 39
sanctions 24, 25, 33, 52, 94, 101, 108, 119, 125, 128, 129
Sandinista government; Sandinista governance 62–64, 66, 68, 70, 72
Sandinistas 57, 62, 63, 64, 67–69
Sandinistas armies; Sandinista military; Sandinista soldiers 63–65, 68, 68, 70
Sandinistas, ex- 66–68, 71
Sandinistas, high-ranking; Sandinista officers 68, 70, 71,
Sapo National Park 49
scarcity theory 38
school drop-outs 44
scorched earth policy; scorched earth strategy 5, 44, 161
secondary forest 79
Secretary of Defense McNamara 151
sectoral reform 108
sects 51
security services 122
self-determination 69, 167
self-development 167
semiautonomous management 170
Senegal 41, 94
settlers 45, 60, 61, 65, 66, 70, 71, 83, 84
shadow economies 99, 100, 109
shadow state 18, 19
shadow taxation 27
shanty towns 43
shipments of arms 121
Siaka Stevens 43
Sierra Leone 2, 19, 35, 37–39, 41, 42, 46–50, 52, 93, 96, 98–100, 110, 117, 131
Sioux 166
slaves; ex-slaves; former American slaves 40, 110
small farmers; small-scale farming 60–63
small loggers 29
smallholders 58, 60, 61
smart sanctions 129
smuggling 11
social movement 18, 96, 103
social networks 82
social relations of production 109
Societe Generale de Surveillance 30
socio-political upheaval 118
soft tissue sarcoma 157
soil erosion 6, 140, 150, 157
soil quality 139, 140
Somalia 94
South Africa 42, 95
South Vietnam 150, 151, 155–157
Southeast Asia 4, 5, 10, 14, 15, 20, 75, 76, 79, 97, 127, 149
Southeast Asia, mainland 75, 76, 79
Southern Africa 97, 98, 102, 111
Soviet Union 63, 96
soybean 58
Spain 100
species loss 141, 142
spices 4

spina bifida 157
spiritual movements 51
sporadic violence 119
spread of disease 9
state actors; state organizations 120, 121
state building 49
state cooperatives 85
state formation in Africa 110
state power 97, 100, 101
state presence 8
state revenues 122
state security forces 121, 123
state-owned companies 79
stillbirths 156
structural adjustment credit 30
structural adjustment programs; structural adjustment 19, 30, 102, 111, 112
subcultures 51
Sub-Saharan African countries; sub-Saharan Africa 7, 98, 102, 115, 123, 128, 131
subsistence needs 75
Sudan 2, 94–96, 98, 99, 159, 162
sulfuric acid 135, 136
sustainable development 13, 15, 18, 60, 105, 115, 132, 141
Sustainable Forest Products Global Alliance 104
Sustainable Management of Cameroon Forests project 108
sustainable utilization 94, 105
swidden cultivation 79, 82, 83, 86, 88, 89

Taliban 141, 144
Tanzania 95, 96, 102
taxation 20, 27, 32, 79, 120, 121
teak 17
technological change 6
technological progress 4
technology 4, 11, 12, 40, 45, 85, 89, 120, 166
technology, capital intensive 89
Thai business 25
Thai forest industry 24
Thai military 23, 25
Thailand 21, 23–26, 30, 31, 80, 83, 129, 10, 161, 164
The Nature Conservancy, TNC 65, 144, 71
timber companies 45, 65, 130
timber exports 25, 28, 29, 129
timber use rights 124
trade boycotts 52
trade networks 118, 122, 124
trade routes 80, 124
traditional livelihood activities 122
trans national corporations;+A960 TNCs 96, 97, 105, 112
transboundary protected area 164, 169
transit countries 130
transnational intervention 103
transnational warfare 95
transparency 33, 108, 111
transportation infrastructure 120, 124
travel bans 38
Treaty of the Prohibition of Chemical Weapons 157
TREES 105
tribal boundaries 165
tribute 79
Trinoxol 155
troop deployment 52
Tropical Forest Action Plan 64, 65
Turkey 100
two-way commodity chain 121

Uganda 2, 94, 95, 98, 99, 101, 103, 121, 161
Ukraine 96
UN arms sanction 13
UN bans on trade of timber, UN ban on Liberian conflict timber 119, 128
UN member states 128
UN military personal 26
UN monitors 25
UN peacekeeping mission 96
UN sanction 25, 52, 125, 128
UN Secretary General 26, 128
UN Security Council sanctions 94, 101
UN Security Council; Security Council 2, 15, 24, 25, 94, 101, 118, 127, 128, 132
UN sponsored elections 28
unequal distribution of wealth 22
unequal resource distribution 38
UNESCO 66, 104
UNICEF 94
Unification Church; Moonies 100
unilateral interventions 52
United Kingdom 100
United Liberian Movement for Democracy; ULIMO 39, 41, 42
United Nations Forum on Forests; UNFF 108
United Nations Mission in Liberia; UNMIL 39, 40, 100, 101
United Nations Mission in Sierra Leone; UNAMSIL 39, 42
United Nations Transitional Authority in Cambodia; UNTAC 24
United States 17, 63, 65, 78, 80–82, 103, 149–152, 154–157

United States military 20, 81, 151, 152, 155, 156
unregulated logging 78, 79, 88, 124
urban educated people 41
urbanization 97
USAID 65, 71, 103, 108
valuable resources 57, 58, 110
veritable frontier 65
vertebrates 141
vested interests 106, 128
Vietnam 23–25, 29–31, 75, 79, 80, 82, 149–158
Vietnam War 10, 149–158
Vietnamese and American forces 75
Vietnamese forces 21, 22, 23, 80, 82
Vietnamese occupying forces 20
Vietnamese troops 17, 23
village history 78
violence, ethnic 50
violence, extreme 2
violence, financing 13
violence, irrational 40
violence, sporadic 119
violence, widespread 10, 11, 61, 142
violence, youth 49
violent plunder 46
violent struggle 38, 47, 51, 95
Violeta Barrios de Chamorro 64, 65, 68, 70, 71
Virunga Volcanoes region; Virunga National Park 161, 162
volatile substances 136
vulnerable group 38, 95

war zone 9, 163
warlords, warlordism 12, 19, 27, 40, 45, 54, 96, 97, 99, 100, 111, 112, 114
warrior cult 50
wartime blockade 3
wartime displacement 10, 77
water catchments 142
water reforms 11
water scarcity 160
wealth accumulation 9
wealth disparities 20
wealthier stakeholders 61
weapons purchases 119
West and Central Africa; West Africa 8, 37–54, 96–99, 101, 102, 104, 114, 127, 132
Westphalian State Form 109
white throated capuchin 164
wild resources 123
wild taro 149
wildlife habitat 73, 87
wildlife hunting 40
wildlife poaching 11, 165
wildlife sanctuaries 29
wood export 79, 100
World Bank 27, 30, 60, 104, 105, 106, 108, 110, 129, 133, 162
World Resource Institute 98, 99, 106, 108
World Summit on Sustainable Development; 1992 Earth Summit 105, 107
World Trade Organization 111
World War II 76, 150
WWF 103, 106

Yaounde Declaration 104
YATAMA 64, 68, 69, 71–74
youngsters, urban semieducated 43
youth violence 49
youth, homeless 43
youth, intellectual 44
youth, Liberia's indigenous 41
youth, rural 43
youth, unemployed 44
youths, Nimba 39

Zaire 13, 37, 161, 171
Zambezi basin 103
Zambia 95, 96, 98, 102, 111
Zimbabwe 95, 101, 102, 113, 121
Zimbabwe Defence Force 104

World Forests (Series)

1. M. Palo and J. Uusivuori (eds.), *World Forests, Society and Environment.* 1999. ISBN 0-7923-5301-3.
2. M. Palo and H. Vanhanen (eds.), *World Forests from Deforestation to Transition?* 2001. ISBN 0-7923-6683-2.
3. M. Palo, J. Uusivuori, and G. Mery, *World Forests, Markets and Policies.* 2001. ISBN 0-7923-7170-4.
4. J.-A. Lamberg, J. Näsi, J. Ojala, and P. Sajasalo (eds.), *The Evolution of Competitive Strategies in Global Forestry Industries.* 2006. ISBN 1-4020-4015-6.
5. W. De Jong, D. Donovan, and Ken-ichi Abe (eds.), *Extreme Conflict and Tropical Forests.* 2007. ISBN 978-1-4020-5461-7.

MIX
Papier aus verantwortungsvollen Quellen
Paper from responsible sources
FSC® C105338

If you have any concerns about our products,
you can contact us on
ProductSafety@springernature.com

In case Publisher is established outside the EU,
the EU authorized representative is:
Springer Nature Customer Service Center GmbH
Europaplatz 3, 69115 Heidelberg, Germany

Printed by Libri Plureos GmbH
in Hamburg, Germany